Nanotemplating

Nanotemplating

Engineering Nanocomposite Materials for the Future

edited by

Humphrey H. P. Yiu

JENNY STANFORD
PUBLISHING

Published by

Jenny Stanford Publishing Pte. Ltd.
101 Thomson Road
#06-01, United Square
Singapore 307591

Email: editorial@jennystanford.com
Web: www.jennystanford.com

British Library Cataloguing-in-Publication Data
A catalogue record for this book is available from the British Library.

Nanotemplating: Engineering Nanocomposite Materials for the Future

ISBN 978-981-4316-92-7 (Hardcover)
ISBN 978-1-003-50999-8 (eBook)

Contents

Preface

Research in nanomaterials has been growing rapidly since the 1990s when scientists learnt more about their formation mechanisms and the relationship between their structures and properties. Using a template to make a nanomaterial allows a good control in quality and specificity. Compared with their bulk cousins, the superior properties seen from nanomaterials are partly due to their high surface area to volume ratio. Such properties have enhanced the efficiency of many reactions and processes. Understanding how these nanomaterials can be made becomes essential for their advancement in future.

Among the nanomaterials commonly found in the literature, nanoparticles are the smallest in dimensions. In Chapter 1, we focus on the synthesis and the applications of these so-called 0-dimensional nanomaterials. There is a wide range of nanoparticles, from inorganic, organic to hybrids. Their design can also be very complex due to the demand on specific properties. Although not all nanoparticle preparation methods require templates, use of a template, notably surfactants, improve the homogeneity of the nanoparticle product.

Use of surfactant templates not only helps nanoparticle synthesis, at a higher concentration, the formation of micelles and liquid crystals triggers the formation of another family of nanomaterials, ordered mesoporous solids. In Chapter 2, we discuss the formation of this unique type of nanomaterials as well as their applications. More interestingly, these ordered nanomaterials can also be used as hard templates themselves to form further nanomaterials. Chapter 3 presents the development of the extended family of ordered mesoporous solids.

There is a limit on how large a micelle can grow and this hampers the expansion of ordered mesoporous solid. In Chapter 4, we learn how to utilize nanoparticles, as introduced in Chapter 1, as the hard templates to form another class of ordered nanomaterials, inverse opals. These nanomaterials with large pores allow applications involving biomacromolecules. More interestingly the optical properties of inverse opal also allow them to be used in optical sensing and photocatalysis.

However, most hard and soft templates introduced in Chapters 1 to 4 only provide physical properties, e.g., template dimensions, for the nanomaterials to build on. In Chapter 5, we introduce the molecular imprinting technology, which allows us to build porous nanomaterials based on the surface chemistry of the templates, enhancing the specificity of the pores, or active sites. Finally, in Chapter 6, we explore the templates found in nature and introduce a series of nanomaterials built from biological templates.

Owing to the demand for high-performance materials and machines, nanomaterials will have to be multifunctional. The challenge is to pack several functionalities on rather limited space but without interference. Understanding the formation and functionalization of various nanomaterials will certainly help to develop the next generation of multifunctional nanomaterials.

Humphrey Yiu

Acknowledgement

First and most importantly, I would express my greatest appreciation to my family, who supported me throughout my study and career in academia. I nearly gave up chemistry twice and it was my family to help me realize that chemistry is what I want to do. I also want to thank my mentors and my colleagues for their inspiration, encouragement, and valuable discussion on various subjects of science. Special thanks go to Prof. David Smith, with whom we have much interesting conversation about all sorts of topics beyond science. Prof. Paul Wright and Prof. Matthew Rosseinsky are also thanked for being such good mentors and I learnt much from them. I also need to thank all my PhD students and postdoctoral researchers. Without their hard work, we would not achieve so much.

Chapter 1

Surfactants and Micelles: Templates for Organic and Inorganic Nanoparticles

Humphrey H. P. Yiu
School of Engineering and Physical Sciences, Heriot-Watt University, Edinburgh, EH14 4AS, UK
h.h.yiu@hw.ac.uk

Nanoparticles are one of the major families of nanomaterials. Indeed, their unique properties can highlight the difference between nanomaterials and their bulk cousins. For instance, nano-sized gold particles are in red to purplish color, in contrast to the usual golden color that we see from bulk gold metal. Bulk Fe_3O_4 (magnetite) is ferromagnetic, while nano-sized (crystal size < 30 nm) Fe_3O_4 is superparamagnetic, i.e., carrying no permanent magnetic moment. Of course, we also have the ever-intriguing quantum dots (or nano-sized semiconductors).

Because of the small size of nanoparticles, they do have some rather seemingly "frightening" properties. For example, nanoparticles are likely to penetrate cell membranes much faster than the micro-sized counterparts. Therefore, the public has the right to worry about the use of nanoparticles. However, many nanoparticles are already widely used in daily life products. Transparent zinc oxide ZnO nanoparticles have been used in sunscreens and medicines for decades, and you may even find

Nanotemplating: Engineering Nanocomposite Materials for the Future
Humphrey H. P. Yiu

ISBN 978-981-4316-92-7 (Hardcover), 978-1-003-50999-8 (eBook)
www.jennystanford.com

titanium dioxide nanoparticles in some chocolate desserts. Nanoparticles are not entirely new technology. Nature has already designed many nanoparticles to improve the lives of some organisms. One interesting example is that nanoparticle-like structure was found on the feet of geckos (Fig. 1.1) and such a nanostructure allows geckos' feet to have super-adhesive properties. Scientists then exploit this property of nanoparticles in the industry for the manufacture of tires for enhanced adhesion to the road surface. Therefore, we should not be scared of nanoparticles but we need to be aware of their property and potentials, as well as potential risks. Instead, we should make use of these delicate little fellows to help us in our everyday lives.

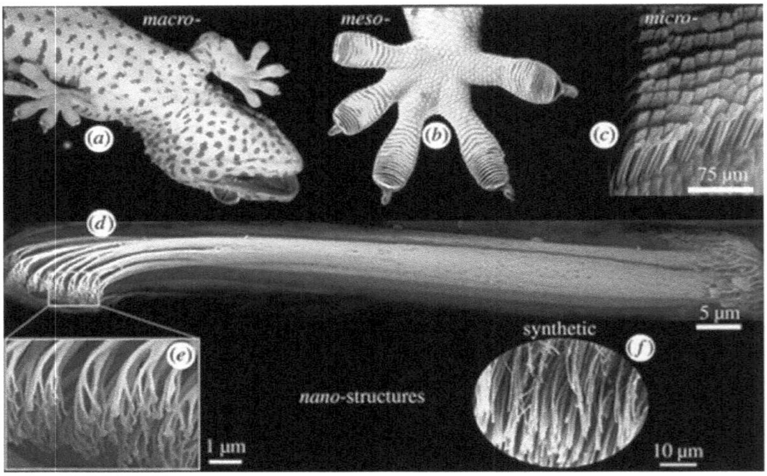

Figure 1.1 Images depicting the nanostructure of the gecko's feet for adhesion (Autumn and Gravish, 2008).

1.1 Introduction

The definition of nanoparticles is very vague and varies from one scientific area to another; a bit like the word "organic," which may mean a vast class of carbon compounds in chemistry or a type of "healthier, natural" food option for the general public. Nanoparticles for biology can be any particulate materials of dimensions below 1 micron. However, for physical scientists, i.e., physicists or chemists, the term "nanoparticles" usually refers to solid particles of a diameter within the range of 5–100 nm. This is because, in this range, many materials exhibit some specific properties that differ from the bulk materials, such as superparamagnetism observed in magnetite Fe_3O_4. Although there are many examples of solid particles of a size below 5 nm, these are often not stable as free particles. They can be stabilized by a support material or in a matrix. For example, many metal particle catalysts are of size below 5 nm, including Pd/C catalyst for hydrogenation, but they can only exist as supported catalysts, on a carbon carrier. Because of their high reactivity, without the support, these small nanoparticles can readily aggregate to form larger particles; this is a process of "sintering." Avoiding sintering is still a challenge in nanoparticle technology.

It is difficult to group nanoparticles into categories because of their huge variety in terms of chemical nature (inorganic vs organic), chemical or physical properties (e.g., soft vs hard), size and shape (e.g., cubic vs spherical). From a chemistry point of view, the simplest classification can be made as *inorganic* and *organic* nanoparticles. Inorganic nanoparticles are mainly crystallites of dimensions in the nanometer scale. These include silica, carbon (e.g., C_{60} fullerene bucky balls), quantum dots, metal oxides (e.g., titania TiO_2 and zinc oxide ZnO), metals and alloys (e.g., PtFe). Organic nanoparticles are commonly made of organic polymers, synthetic or natural. These organic nanoparticles can also be made with copolymers or blended polymers in order to optimize their properties.

There are many methods to make nanoparticles, depending on the required morphology and size. The choice of methods is also dictated by the end use of the prepared nanoparticles. We can vaguely group these methods into "top-down" and "bottom-up." Top-down methods mean that the nanoparticles are prepared from a larger bulk material by "cutting" down to nano-sized particles. For example, ball milling is widely used in engineering and materials science to "grind" down a piece of bulk material into smaller fragments. Nowadays, ball milling technology can achieve producing submicron fragments (or several 100 nm), towards nanoscaled materials. Ball milling is a simple method to produce near-nanoscaled materials in large quantities. However, the quality of the ground materials is usually coarse with a rough surface and a random morphology. For most high-end, advanced applications, in particular biomedicine, this is just not good enough. Moreover, during the ball milling process, high heat is to be generated due to friction and therefore, it is not suitable for thermally unstable materials, including polymers.

A bottom-up approach is to build the nanoparticles from small units, which are usually free ions or individual molecules. Controlling the growth of particles is key to a successful synthesis but small particles tend to grow larger due to their instability. Stabilizing reagents can be added to inhibit the growth of particles to within nanoscale dimensions. Another way to achieve this is to crystallize the particles very quickly so that there is no time for growth. This theory is commonly used in the template-free precipitation method. The final method is setting out a nanoscaled boundary. This is the most sophisticated method with high precision in controlling the dimension of the final product.

1.2 Surfactants and Micelles

Despite their technical-sounding name, we use surfactants and make micelles every day. And we all have some form of surfactant at home. Take a look at your shampoo, washing-up liquid, washing powder and other household detergents; they all have an active ingredient called surfactant. It can be cationic, anionic or even non-ionic surfactants, all serving the same purpose: forming micelles to remove oily dirt. Also, the surfactant is nothing new. We have been using surfactants for cleaning for centuries. However, here, we are not talking about cleaning. We are talking about making nanomaterials. So how do surfactants act as templates for making nanomaterials? To answer this, we need to start with micelles formation.

Surfactants have a hydrophilic (water-like) and a hydrophobic (water-hate) part in their molecules. They are usually soluble in both water and oil. At a certain concentration, these surfactant molecules start to self-assemble and form spherical structures called micelles. This concentration is called *critical micelle concentration* 1 or CMC 1, which is characteristic for every surfactant. In an aqueous medium, these micelles have a hydrophobic core and a hydrophilic shell. This structure is reversed when the micelles are formed in an oily medium, and usually these micelles are called reverse micelles or inverse micelles. Figure 1.2 illustrates the formation of micelles and reverse micelles in an aqueous and oily medium, respectively.

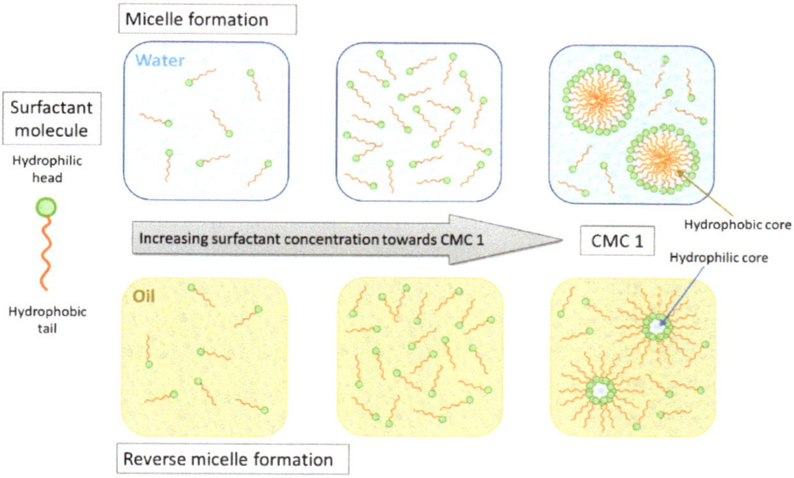

Figure 1.2 Illustration of the formation mechanism for micelles and reverse micelles.

The size of micelles is variable, from several nanometers to a few hundred nanometers, depending on the concentration of surfactants, temperature, pH of the solution, and the presence of swelling agents. However, micelles usually are of the same size throughout one homogeneous solution. This is an essential property to make nanoparticles of homogeneous size. The size of these micelles can be examined using dynamic laser scattering, which is now a very common technique to study the size distribution of nanoparticles and microspheres.

1.3 Nanoparticles Templated by Micelles and Surfactants

In this section, we will look at several examples of nanoparticle preparation procedures aided by micelle or surfactant templating.

1.3.1 Microemulsion for Preparing Polymer Latex Nanoparticles

Synthetic polymers, or plastics to laypersons, are found everywhere in our homes and we use some of these every day, in all sorts of forms and shapes. For example, most plastic bottles for soft drinks or food are made from polypropylene. Polyethylene is used as cling films and carrier bags, and everybody has heard of polystyrene cups and plates. These polymers are mostly made with a polymerization process of monomer molecules (e.g., ethylene or styrene) initiated by free radicals. One major difficulty of this process is to control the growth of polymer, in particular at a molecular level. Therefore, to make nanoparticles with these polymer materials requires some form of physical restriction.

Since most of the monomers for making these polymer materials are oil-soluble organics, the use of micelles to control their growth is ideal. In a typical reaction, a surfactant solution is prepared, followed by adding monomers and initiators into the solution. Because of their hydrophobic nature, these monomers and initiators will get inside the cores of micelles and swell them. Sometimes, this mixture is also referred to as a microemulsion. The polymerization process starts when the temperature is increased and free radicals start to form from the initiators. Constant, fast agitation is required to ensure the stability of the emulsion. Once the polymerization process finishes, these nanoparticles are quite stable at room conditions. The size of these nanoparticles ranges from 100 to 500 nm. As previously mentioned, one major advantage of this process is the homogeneity of the nanoparticles produced. The size distribution is very narrow in a single pot. Indeed, these nanoparticles are sometimes used as standards for size determination for other nanomaterials. On the other hand, the limitation of this process is also their size. Micelles also have their limit and, together with the monomers, usually they cannot be smaller than 100 nm. Although there are reports on the template-free synthesis of these latex particles, the quality of the resultant latex is usually lower due to the lack of size control provided by the template. Hence, a wider range of particle sizes is likely to be observed.

Box 1.1

Preparation of polystyrene latex nanoparticles with the micelle-templating procedure

Extreme caution has to be taken since most quantum dot materials and their precursors are highly toxic. This experiment must only be carried out by experienced synthetic inorganic chemists, with full risk assessment, COSHH and precautions in place.

There are many protocols for making polystyrene latex particles using the micelle-templating route. The method presented is just one example, using sodium dodecyl sulfonate (SDS) as the surfactant template (Li et al., 1997).

Sodium dodecyl sulfonate (0.18 g) is mixed in 50 mL of distilled water in a 100 mL two-necked flask equipped with a mechanical stirrer and a nitrogen inlet tube. The solution is stirred at 60°C under nitrogen until all the sodium dodecyl sulfonate is dissolved.

Purified styrene (15 mL) is added to the surfactant solution dropwise, and the emulsion is stirred vigorously for 90 min until a stable emulsion is formed.

The initiator, potassium persulfate (0.1 g), is dissolved in 1 mL of water, then added slowly to the emulsion with stirring. The reaction temperature is maintained at 80°C and the emulsion is allowed to react for 24 h under nitrogen.

A stable white emulsion of polystyrene latices is obtained at the end of the reaction.

The polymer can be isolated from the emulsion for characterization purposes. To do this, 10 mL of the emulsion is diluted with 10 mL of water in a 100 mL beaker and precipitated by slowly adding a saturated $KAl(SO_4)_2$ electrolyte solution with gentle stirring. The white precipitate is filtered and dissolved in THF, followed by reprecipitation in a large quantity of hot water, then filtered again and washed with methanol to remove the surfactant. The product can be dried at 50°C at a reduced pressure. Characterization can be carried out using FTIR, NMR and dynamic light scattering DLS.

1.3.2 Reverse Micelles for Preparing Magnetic Nanoparticles

As we have discussed, polymer nanoparticles can be prepared with micelles because of the hydrophobic nature of the monomers and initiators. However, what can we do if these precursors are water-soluble instead? We can simply reverse the phase and use reverse micelle templating for making these nanoparticles.

Preparing a solution of reverse micelles is simple (see Fig. 1.3). First, a certain amount of the chosen surfactant is to be dissolved in a non-polar solvent (oil phase) such as hexane, followed by adding a small amount of water. With constant, high-speed stirring, water will be trapped inside the reverse micelles and an emulsion is formed. To make inorganic nanoparticles, the inorganic precursors need to be first dissolved in water and added to the emulsion. The formation of nanoparticles inside the reverse micelles is usually triggered by a change in pH, e.g., most metal oxides form in an alkaline medium. This method is particularly suitable for preparing metal oxide nanoparticles because many of the metal oxide precursors (e.g., chlorides and nitrates) are water soluble.

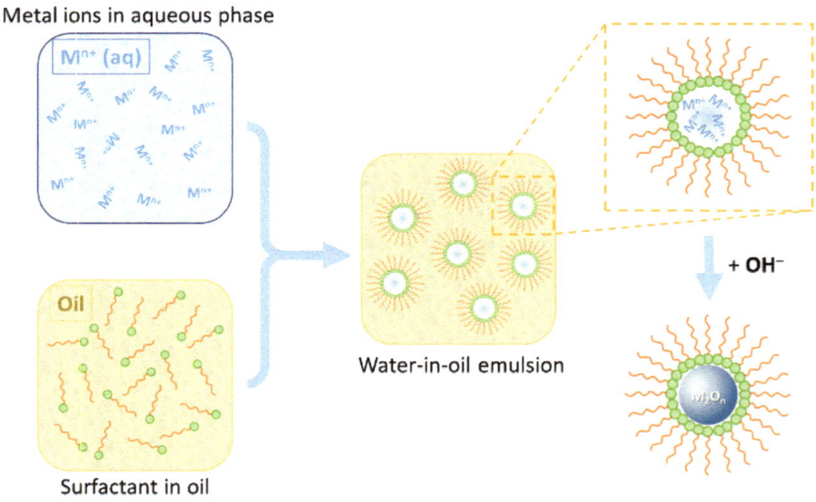

Figure 1.3 Schematic illustration for the preparation of metal oxide nanoparticles via the reverse micelle route.

One example of using this reverse micelle route to prepare nanomaterials is the formation of the magnetic silica-iron oxide nanocomposite particles. The size of these nanoparticles can be tuned by varying the amount and nature of the surfactant used, the amount of water, and the temperature of the reaction. These nanoparticles have many uses in biomedical research due to their magnetic property, which leads to applications such as cell sorting, isolation and purification of biomolecules including DNA, proteins and antibodies (Yiu, 2011).

Despite the good quality of nanoparticles, the use of reverse micelles for this purpose is not popular with scientists. The large amount of oil or organic solvent required, and to be removed eventually, can be problematic. As a result, scaling up is difficult and so it is unlikely to be developed into an industrial process. However, for scientific research, reverse micelle templating is still a useful process.

1.3.3 Solvent-Free Thermal Decomposition in Oleic Acid and Oleylamine

To make metal oxide nanoparticles smaller than 10 nm, the decomposition of metal precursors at high temperatures in the presence of oleic acid and oleylamine is commonly used (Mourdikoudis et al., 2022). Oleic acid and oleylamine are oily surfactants with high boiling points (ca. 360°C), which is higher than the decomposition temperature of some metal-organic complexes (ca. 250°C) such as iron(III) acetylacetonate, or $Fe(acac)_3$. The surfactant prevents the growth of the nanoparticle, similar to the reverse-micelles route. However, this is technically not micelle-formation as sometimes there is only one phase in this method. Also, the surfactant chosen should be able to form a strong attractive force with the transition metal. Hence, surfactants with carboxylic acid groups, which form complexes with most transition metals, are to be used in this method. Furthermore, these surfactants must have a high boiling point, higher than the decomposition temperature of the metal precursors. These factors limit the choice of surfactants to be used in this method.

While still protected by the surfactant template, the prepared nanoparticles can be modified by simple reactions such as reduction. For example, oleic acid-templated cobalt oxide nanoparticles can be reduced using $NaBH_4$ and form metallic cobalt nanoparticles (Pirkkalainen et al., 2008). These cobalt nanoparticles are also in different morphologies, depending on the reaction conditions. Nanoparticles with cubic, rod, and even star shapes are found. This is due to the early stage of crystallization. Unfortunately, this method is not applicable to some other nanoparticles prepared with the oleic acid templating method. For example, metallic iron nanoparticles are not formed using this reduction method because these nanoparticles are just too unstable and easily re-oxidized in water (or going rusty).

The major problem of using this high-temperature process in making nanoparticles is their small scale. The quantity of nanoparticles produced in every batch is a few hundred milligrams at best. However, this process provides small-size nanoparticles (down to 5 nm) with great homogeneity. Hence, it is still commonly used in research. Another problem is "phase transfer," which means transferring materials from an oily phase to an aqueous phase. This is due to the strong binding between the surfactants and metal. Since these nanoparticles are only "oil-soluble," ligand exchange is usually required for an effective transfer into a water medium, which is preferred by most biological applications. Poor ligand exchange will result in the agglomeration of nanoparticles and loss of their unique nanoparticle properties.

1.3.4 Quantum Dots

Many of us have heard of quantum dots but some still do not have much idea what they are. Since quantum dots have their unique luminescent property that no other materials can match, we ought to have some idea of what they actually are and how they are made. Such information will help us to design systems suitable for human needs, and fully utilize these fantastic nanomaterials.

In simplified terms, quantum dots are nanoparticles of some semiconducting materials. The particle size of these nanoparticles is between 3–10 nm (Wang et al., 2007a). They have an exceptional luminescent property, which is somewhat 50 times brighter than conventional fluorescent dyes. Moreover, the excitation/emission wavelength (or simply their colors) is not governed by their molecular structure like other organic fluorescent dyes. Their luminescent property is largely governed by their particle size, and usually a longer excitation/emission wavelength is found in quantum dots of larger particle size. This is the reason why they are less vulnerable to bleaching. Unfortunately, there is no simple formula to predict the excitation/emission wavelength of quantum dots and different semiconducting materials have a different range of excitation/emission. Figure 1.4 shows the emission wavelength of several common semiconducting materials. As we can see, among these materials, CdSe covers the majority part of the visible spectrum and therefore, CdSe quantum dots are the most commonly used.

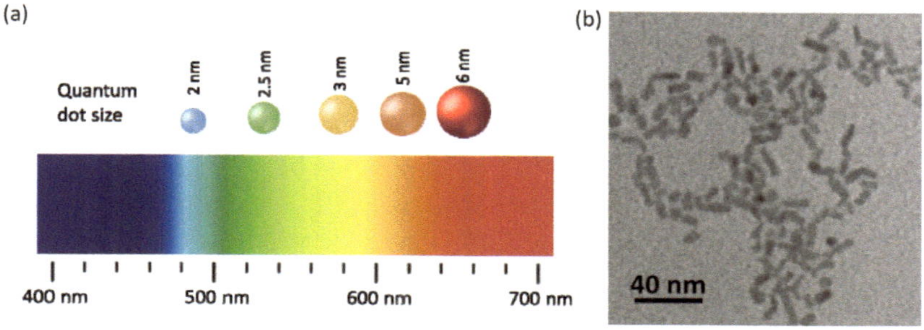

Figure 1.4 (a) The relation between the emission wavelength (excitation 450 nm) and the size of CdSe quantum dots, (b) TEM image of a CdSe quantum dot sample.

Since the size of the quantum dots is critical to their luminescent property, preparing semiconducting nanoparticles of a narrow size range is essential to the success of these materials. Until the mid-1990s, scientists first reported the use of phosphine and phosphine oxide surfactant to make CdSe nanoparticles and observed their unique luminescent property (Nirmal et al., 1993). There are several similarities between the use of phosphine (e.g., trioctylphosphine, TOP)/phosphine oxide (e.g., trioctylphosphine oxide, TOPO) surfactants to make quantum dots and that of oleic acid for transition metal oxides. First, these phosphine and phosphine oxide surfactants are used to control the growth of semiconducting nanoparticles by keeping them apart to avoid sintering. Second, their high boiling point is essential as the decomposition of precursors occurs at around 300°C. Finally, the functional groups (P and P=O) form a strong interaction with the elements (e.g., Cd and Se) of quantum dots.

Box 1.2

Preparation of CdSe Quantum dots

Extreme caution has to be taken since most quantum dot materials and their precursors are highly toxic. This experiment must only be carried out by experienced synthetic inorganic chemists, with full risk assessment, COSHH and precautions in place. The reaction is mainly carried out in an inert atmosphere such as argon or nitrogen.

The typical preparation of TOP/TOPO capped CdSe nanocrystallites follows:

1. Solution A is prepared by adding 1 mL (13.35 mmol) of methyl cadmium (Me_2Cd) to 25 mL of TOP in the drybox.
2. Solution B is prepared by adding 10 mL of the 1 M TOPSe stock solution (10 mmol) to 15 mL of TOP.
3. TOPO (50 g) is dried and degassed in the reaction vessel by heating to 200°C at 1 atm for 20 min, flushing periodically with argon.
4. The temperature of the reaction flask is then stabilized at 300°C under 1 atm of argon.
5. Solutions A and B are combined and loaded into a 50 mL syringe in the drybox. The heat is removed from the reaction vessel. The syringe containing the reagent mixture is quickly removed from the drybox and its contents are delivered to the vigorously stirring reaction flask in a single injection through a rubber septum.
6. The rapid introduction of the reagent mixture produces a deep yellow/orange solution with an absorption feature at 440–460 nm. This is also accompanied by a sudden decrease in temperature to 180°C. Heating is restored to the reaction flask and the temperature is gradually raised to 230–260°C.
7. Aliquots of the reaction solution are to be sampled at regular intervals (5–10 min) and absorption spectra are taken to monitor the growth of the crystallites. The best quality samples are prepared over a period of a few hours of steady growth by modulating the growth temperature in response to changes in the size distribution as estimated from the absorption spectra. The temperature is lowered in response to a spreading of the size distribution and increased when growth appears to stop.

8. When the desired absorption characteristics are observed, a portion of the growth solution is transferred by cannula and stored in a vial. In this way, a series of sizes ranging from 1.5 to 11.5 nm in diameter can be isolated from a single preparation.

Because of their small size and high surface area, many quantum dots are unstable in room conditions without protecting coatings. The phosphine/phosphine oxide surfactants are able to stabilize the quantum dots but they are only oil soluble. For many applications, particularly biomedical applications, water-soluble quantum dots are essential. Such a phase transfer problem is still one of the obstacles in the development of quantum dots. One common method is to coat CdSe quantum dots with a ZnS shell and further coated with functional ligands. ZnS is also relatively less toxic and does not interfere with the luminescent property of the CdSe core. Ligands, usually with thiol groups for a strong association with ZnS, can not only make the quantum dots water-soluble but also provide necessary functional groups for further specific functionalization.

1.3.5 Surfactant-Templated Gold Nanoparticles and Nanorods

Gold nanomaterials have been very popular in many applications due to their easy functionalization with thiol groups. As a result, much effort has been made to develop gold nanomaterials with various morphologies. Surfactant templating has been one of these methods for making gold nanoparticles (Au NPs) and gold nanorods. For example, Spirin et al. demonstrated using a binary surfactant mixture of cetyltrimethylammonium bromide (CTAB) and benzyldimethyltetradecyl ammonium chloride (BDTAC) can control the morphology of gold, from nanoparticles to nanorods. (Spirin et al., 2020). However, many reports on Au NPs are based on non-template routes since the removal of surfactant templates can be problematic. Also, these Au NPs from non-template routes are of good quality (e.g., size distribution and crystallinity) for the desired applications.

Similar to quantum dots, gold nanomaterials show unique characteristics due to their nano size. For example, gold nanoparticles are not in shiny yellow metallic color. Instead, their color can be red, purple or blue, depending on the particle size (Duan et al., 2014). Such property had been exploited for use in detection as a biomarker. Gold also had one property that makes it highly sought after as a sensing material: being plasmonic. The plasmonic resonance property allows gold nanoparticles to enhance electromagnetic signals for sensing and detection of molecules (e.g., pollutants) at ultra-low concentrations. (Yan, 2013).

One obvious disadvantage of using gold nanomaterials is their cost. Gold has been known as a precious metal for thousands of years and its price has been rising in recent years. Products involving gold components are usually thought to be expensive. However, compared with many specific biomolecules, gold nanoparticles are relatively cheap. For example, in recent years, the lateral flow test kits for Covid-19 are made with gold nanoparticles conjugated with antigens (Huang et al., 2020) and hence the indicator lines were reddish-purple in color. These tests are relatively cheap (<$10 per test) and widely available.

1.4 Characterization of Nanoparticles

1.4.1 Dynamic Laser Scattering (DLS)

To characterize these nanoparticles, in particular their size distribution, dynamic laser scattering, or DLS, is the most common technique. Dynamic light scattering (also known as photon correlation spectroscopy, or PCS) allows the determination of the size of particles in the submicron to nanometer range in a suspension. These particles are found in Brownian motion in suspension and the diffusion speeds for this are inversely proportional to the dimension d_p (hydrodynamic diameter) of the particles, D is the diffusion constant:

$$D = \frac{kT}{3\pi\eta d_p} \tag{1.1}$$

where k is Boltzmann's constant, T is temperature, and η is the viscosity of the liquid.

As the equation suggests, these are important parameters, which must be accurately known for the measuring of particle sizes. Because of the Brownian motion of the particles, the backscattered light gets a frequency shift (Doppler shift). The resulting intensity variation is recorded by a highly sensitive detector.

Since many conventional analyzers mathematically resolve the time dependency of the intensity variation by auto-correlation, the term photon correlation spectroscopy (PCS) has also become commonly used for this method. With these devices, the detection of the scattered light is made at a 90° angle to the incident laser light (e.g., 675 nm red light). However, the use of such analyzers is restricted to very dilute dispersions.

This method has several advantages: first of all, the experiment duration is short (several minutes) and it is almost all automatized so that for routine measurements extensive experience is not required. Moreover, this method has modest development costs. The advantage of using DLS is the potential to analyze samples with broad distributions of species and wide molecular mass range (e.g., a native protein and various sizes of aggregates), and to detect very small amounts of the higher mass species (<0.01% in many cases). Furthermore, with this technique it is also possible to obtain absolute measurements of several parameters of interest, like molecular weight, radius of gyration, translational diffusion constant, etc.

Dynamic Laser Scattering has some limitations. For example, since it is based on the scattering of light, DLS gives high inaccuracy for the analysis of dark particles, which absorb light instead of scatter. The model used for calculating the particle size is based on non-absorptive particles. Magnetite and carbon nanoparticles are good examples of dark particle (mostly black). Also, DLS measures the "hydrodynamic" size, not the actual size of the particles. They could be of rather different values.

1.4.2 Scanning Electronic Microscopy (SEM)

If we want to prove a material is of "nanoscale" dimensions, the most convincing method is examining the photographs or images of the nanomaterial. A standard optical microscope has a limitation in its resolution at around the range of visible wavelengths (ca. 400–700 nm), with a magnification of ×400 to ×1000. This is not enough for us to "see" nanomaterials.

The invention of electron microscopy broke the resolution barriers associated with visible light. Among electron microscopes, scanning electron microscopes (SEM) are commonly used to study materials down to a "submicron" scale. In general, SEM uses a beam of electrons, with an accelerating voltage at around 5–20 kV, focusing on the solid sample surface to generate a variety of signals, e.g., secondary electrons, backscattered electrons (BSE), diffracted backscattered electrons, photons, visible lights as well as heat. These signals allow studies on texture/morphology, elemental composition, crystalline structure and orientation of the components in the sample.

For most SEM analyses, the samples are "adhered" onto aluminum sample stubs using carbon-based tapes. A thin layer of gold (or sometimes palladium) coating on the samples is then coated to ensure sample conductivity. If the electron beam is aligned at a 90° orientation to the sample, a 2D image is generated, showing spatial variations in texture or morphology on sample surfaces. The areas of interest can vary from ca. 10 mm down to 5 μm in width, using a scanning mode. Typically, an SEM can achieve a magnification of up to ×30,000 with a spatial resolution of <20 nm. As a result, it may not be good enough for nanostructures much smaller than 100 nm. If the sample is tilted away from 90° to the incident electron beam, "pseudo 3D" images can be generated, showing some depth and texture of the sample. Moreover, SEM can also perform useful analyses on selected point locations on a sample surface; this approach allows qualitative or semi-quantitative analysis on elemental compositions (using EDX or EDS mode), crystalline structure, and crystal orientations (using EBSD). It is worth noting that SEM images should be black-and-white. We should note that colored SEM images shown in some articles or magazines, especially for biological samples such as cancer cells, bacteria or viruses, are artificially colored as shown in Fig. 1.5.

In SEM analyses for many nanomaterials, EDX elemental analyses are particularly useful to confirm the identity of the samples. In an EDX mode, the photon signals generated in the form of characteristic x-rays are used. Depending on the resolution of the SEM unit, elemental mapping analysis can be carried out from a detailed map for the distribution of elements generated. Indeed, such a function is essential for studying heterogeneous samples.

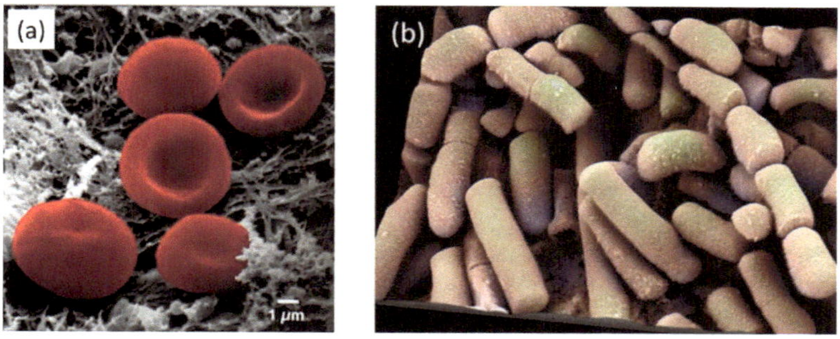

Figure 1.5 SEM images of cells with artificial coloring. (a) red blood cells, and (b) bacteria *Bacillus cereus*. (Mignot, 2018).

1.4.3 Transmission Electron Microscopy (TEM)

Although SEM is a powerful technique allowing imaging of samples at a resolution down to <20 nm, many nano-scaled samples, e.g., quantum dots and gold nanoparticles, are of much smaller dimensions (< 20 nm). At this dimension range, SEM is simply not good enough to provide detailed morphology information. Transmission electron microscopy (TEM) is therefore used instead. Similar to SEM, TEM is a microscopy technique using a beam of electrons but, unlike SEM, the beam is *transmitted* through a specimen to form an image. Usually, a TEM sample is mounted (or deposited) on a TEM specimen grid, which is commonly made of copper. The sample needs to be less than 100 nm thick to allow electrons to go through. When the electron beam is "going" through the sample, there are interactions between the electron beam and the sample leading to generating an image. Digitized TEM images are commonly used nowadays. However, imaging using plastic films (negatives) and printing images on photographic papers were standard before the year 2000.

Compared with SEM, TEM allows imaging of a sample at a high resolution due to the wavelength of the electron beam used is much smaller. As a result, a sub-nanometer resolution is common for a standard TEM unit. For some more advanced units, TEM can even achieve a sub-Ångström (0.1 nm) level of resolution, e.g., scanning TEM or STEM, which allows capturing images of atoms and lattices of samples with fine details. Therefore, TEM is considered as a powerful analytical tool in many scientific areas in physics, chemistry and biology. In particular, in niche areas such as nanotechnology and advanced materials, TEM has become a "must-have" item.

There are a wide variety of TEM instruments operating at many modes for a wide range of advanced applications. For standard imaging purposes, there are a number of "contrasting mechanisms" including contrast generated due to (i) density or thickness of the sample (mass-thickness contrast), (ii) electron density of different elements or their atomic number (Z contrast, e.g., carbon vs gold), (iii) shifts in phase (phase contrast), (iv) variation in crystallographic structure (crystallographic contrast), etc. All these mechanisms are specific for certain types of information, allowing users to have detailed study not only on the morphology but also on the structural chemistry of the sample at a nanoscale, or even at an atomic level. As a result, TEM is regarded as an essential tool for studying nanomaterials as well as other fields including biology and bulk materials.

Table 1.1 Various operating modes and image contrast mechanisms to be used in TEM

Operating mode	**Applications**
Conventional TEM	Standard imaging, materials, biological samples
Scanning (S)TEM	Advanced materials at nanoscale
Electron diffraction	Identification of materials and study on crystallinity
EDX	Elemental analysis of materials
Contrast mechanism	**Applications**
Amplitude: Mass-thickness contrast	Imaging non-crystalline and amorphous samples
Amplitude: Diffraction contrast	Imaging for defects on crystalline samples
Phase contrast	Imaging for crystalline samples

1.4.4 Atomic Force Microscopy (AFM)

Atomic force microscopy (AFM) is an advanced imaging technique with resolution down to a sub-nanometer level, much higher than many optical methods. Unlike other microscopic methods, AFM generates images based on the mechanical force between the sample and the probe (or the cantilever), similar to "feeling" through the sample surface without looking at it! Such a high resolution was due to the lack of optical lenses and light (or beam) source. Also, sample preparation is relatively simple and AFM can also be operated at atmospheric pressure. Mechanically fragile samples can therefore be studied.

The most important part of an AFM unit is the flexible cantilever which is fitted with a Si or Si_3N_4 sharp tip (of several nanometers) at the end. The main function of this tip is to "feel" or "touch" the sample. When the tip is being brought close to the sample surface, the forces between the sample surface and the tip deflects the cantilever, generating a signal. Such deflection can also be monitored using a laser and the signal generated is fed to the main electronics. Such action can be programmed to scan the surface of a sample in an x–y grid. Together with the "depth" information (or z), a 3-D image can be generated, giving a map of topography (see Fig. 1.6). Resolution generated from such action can be very high because the size of the tip is of only a few nanometers.

There are several modes of imaging in AFM: contact mode, tapping mode and non-contact mode. The choice of imaging mode depends on the nature of the sample and the information required to be generated. However, the area of imaging (or scan) is limited to ca. 100 μm × 100 μm, which is nonetheless enough for most nanomaterials. One powerful feature of AFM is that the tip can be tagged with functional molecules or biomolecules to induce specific interaction between the tip and the sample surface. Therefore AFM can be used to study some specific surface chemistry, which is less likely for other microscopic methods such as SEM and TEM.

Figure 1.6 illustrates how an AFM can be used to image nanoparticles. When a sample of nanoparticles is scattered on the sample holder (usually a piece of thin mica), the nanoparticles will induce some "mounts," which can be shown on a 3-D topography map. A cross-section of the 3-D contours can also be generated and the size of the nanoparticles can therefore be measured.

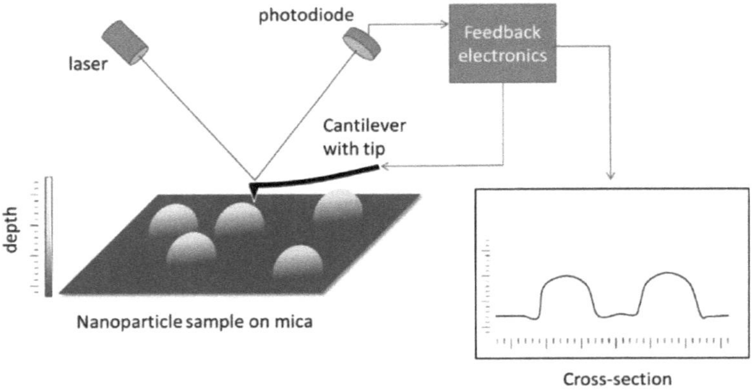

Figure 1.6 Illustration of the principle of AFM in generating a 3-D topography map at a nanoscale.

1.4.5 Powder X-Ray Diffraction (XRD) and Sherrer Analysis

Powder x-ray diffraction (XRD) is a common technique for studying and identifying the structure of a crystalline material. To describe XRD in simple terms, when a monochromatic beam of x-ray hits the surface of a crystal at an angle θ, every atom in the crystal scatters the x-ray beam in all directions. Constructive interference of reflected x-ray radiation only takes place if the arrangement of atoms is regular and repetitive. The intensity of the reflected x-ray beam depends on the atoms located at each plane of the crystal and the location of the atoms. Atoms with more electrons give reflected beams with higher intensities. The interference of such reflected beams gives a specific diffraction pattern for the sample. Figure 1.7 shows the simplified diffractions of an x-ray beam from a set of crystal planes. Intense reflections only occur at certain values of θ, depending on the wavelength (λ) of incident x-rays and the spacing between planes in the crystal lattice (d), or simply d-spacing, expressed by the Bragg equation below:

$$n\lambda = 2d\sin\theta, \quad (1.2)$$

where $n = 1, 2, 3...$

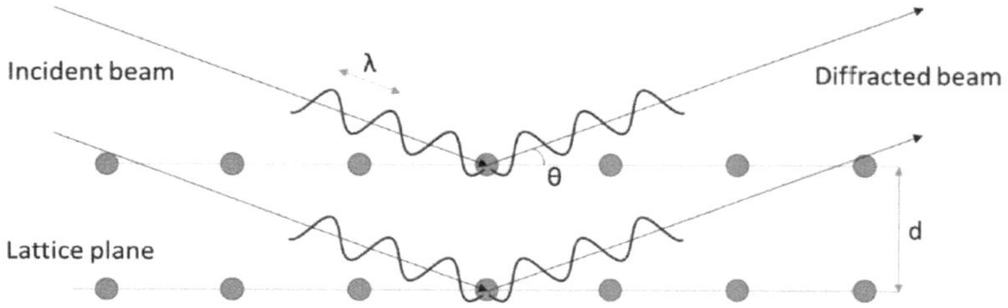

Figure 1.7 Illustration for XRD when a beam of x-rays hits and diffracts from a set of crystal plans.

Powder XRD is the simplest technique to obtain x-ray diffraction patterns, first introduced by Debye and Scherrer. In powder XRD a finely divided polycrystalline powder with random orientations of crystallites is used instead of a single crystal (as in single crystal XRD). As the powdered sample contains a large number of crystallites at random orientations, there will be some at the Bragg reflection angle from each set of planes. When the x-ray beam with known wavelength λ is at an incident angle θ, the reflected beam from the powdered sample makes an angle of 2θ with the incident beam. The sample and the detector are rotated during the measurement. The intensity of the reflected beam is then measured by the detector and recorded. Reflections from sets of planes of the powdered sample interfere constructively when the Bragg condition is met and lead to high intensities measured at the detector. The set of d-spacing data can be obtained from the diffraction pattern based on the reflection intensity against 2θ, which can be converted to d-spacing using the Bragg equation. As a result, different materials will generate their characteristic diffraction patterns according to their lattice structure.

However, when the diffraction patterns are compared between a bulk sample and a sample of nano-sized crystals (crystallites), the diffraction peaks become broader. Such a broadening effect can be used to estimate the size of the crystals on a nanometer scale. This is usually referred to as the "shape factor" in XRD, and a specific equation, derived by Paul Scherrer in 1918, is commonly used for

$$\tau = \frac{K\lambda}{\beta\cos\theta}, \tag{1.3}$$

where K is the shape factor, λ is the x-ray wavelength, β is the line broadening at half the maximum intensity (FWHM) in radians, and θ is the Bragg angle; τ is the mean size of the ordered (crystalline) domains, which may be smaller or equal to the grain size (Scherrer, 1918; Holzwarth and Gibson, 2011). The dimensionless shape factor has a typical value of about 0.9, but it varies with the actual shape of the crystallite. The Scherrer equation is limited to nanoscale particles. It is not applicable to grains larger than about 0.1 μm, which precludes those observed in most metallographic and ceramographic microstructures.

It is important to realize that the Scherrer formula provides an estimate on the particle size. The reason for this is that a variety of factors can contribute to the width of a diffraction peak; besides crystallite size, the most important of these are usually inhomogeneous strain and instrumental effects. If all of these other contributions to the peak width were zero, then the peak width would be determined solely by the crystallite size and the Scherrer formula would apply. If the other contributions to the width are non-zero, then the crystallite size can be larger than that predicted by the Scherrer formula, with the "extra" peak width coming from the other factors.

Since the "shape factor" (K) is dominating the equation above, it is important to choose the correct shape factor for accuracy. Although in most analyses, we take $K = 0.9$ for spherical particles, this can be incorrect as many nanoparticles are non-spherical. In some cases, the shape of particles in one single sample batch can be heterogeneous too. Nonetheless, Scherrer analysis does provide us a good number as a guideline for comparison.

The merit of X-ray diffraction is twofold; XRD not only provides the chemical identity of the nanoparticle samples but also indicates the particle size via Scherrer analysis. In terms of limitation, XRD is a technique based on the crystallinity of the sample, its capability becomes limited for studying amorphous samples, and many nanoparticles are amorphous (e.g., polymer nanoparticles). It also does not give much information about coatings. Therefore, complementary techniques (e.g., FTIR or Raman spectroscopy) are still required.

1.5 Nanoparticles of Higher Complexity

We have just learned that high-quality, homogeneous nanoparticles can be made through micelle templating or surfactant-aided synthesis. Unfortunately, most of these nanomaterials are so-called dimensionless or 0-D as they are simply tiny spheres with a simple chemical composition. Gold nanorods are one of the few exceptions. Indeed, this is the main limitation of these simple self-assembled templates. Systems with higher complexity can be too unstable to form the target nanomaterials.

For nanomaterials with multi-components or higher complexity in structure, multi-steps of post-synthesis functionalization or assembly are required. Even so, such assembly is still in early development. For example, gold and silver nanoparticles are successfully assembled into a number of nano-patterned materials (Ruan et al., 2014). However, these assembling methods tend to be non-universal and non-transferable to other systems. Hence, specific systems are necessary to be developed for certain target materials. Luckily, the scientific communities are working hard on investigating new technology and templating systems for nanomaterials with higher complexity are being reported.

1.6 Surface Functionalization of Inorganic Nanoparticles

Many people are talking about how magnificent nanoparticles are. However, quite often it is the surface functionalities on the nanoparticles carrying out the task chemically and biologically while the cores of the nanoparticles are only providing some physical properties to supplement the chemical and physical properties on surface. Sometimes, the nanoparticle core has no particular function except being a platform for building functional groups on the surface of nanoparticles. For example, silica nanoparticles are rather inert and provide no significant physical property for any task. Gold nanoparticles are also inert without functionality. Other examples include many polymer nanoparticles. One example is polystyrene nanoparticles. Spherotech, an American technology company, sells its functionalized magnetic nanoparticles for biological applications. These particles are polystyrene-based particles with iron oxide nanoparticles attached on the surface to provide magnetic properties, but there is a wide range of surface functionalities to offer applications in different areas.

To functionalize the surface of nanoparticles, we have to understand the surface chemistry of various types of nanoparticles. One method may be suitable for one type of nanoparticles but unlikely to be useful for the others, although there are some exceptions. Also different functionalization methods usually give functional groups with different stabilities. Therefore, it would be easier for us to discuss functionalization according to the nanoparticle types.

1.6.1 Functionalization of Gold Nanoparticles

Gold nanoparticles are among the most popular nanoparticles used in bionanotechnology. One major reason is that they are easy to be functionalized, far easier than most other nanoparticles. First of all, gold is a stable element and it can withstand high acidity and basicity. More importantly, gold has a strong affinity to thiol groups. Therefore, any molecule with a free –SH group can bind strongly onto the gold surface without any coupling reagent. This property has been well-exploited in surface sciences when scientists use a flat gold surface (Au-111, smooth at an atomic level) for studying interaction forces between molecules (Terán Arce et al., 1997).

One of the earliest, also simplest, examples is the use of alkylthiols (R-SH) to functionalize gold nanoparticles (Badia et al., 1996). Although alkylthiols do not have any specific functionality, they stabilize the gold nanoparticles by building a protective layer around the nanoparticles and prevent them from aggregation. These alkylthiol-functionalized gold nanoparticles also become "oil soluble" due to the hydrophobic shell. Common solvents for these nanoparticles include toluene and n-hexane. This also opens up many opportunities for functionalization of higher complexity. For example, thiolated carboxylic acids (HS-R-COOH) are commonly used for the functionalization of gold nanoparticles and make them "water-soluble" because of the hydrophilic property of –COOH groups (Sendroiu et al., 2006). Gold has a much higher affinity to –SH than –COOH groups. Therefore, no HS-R-COOH molecule will be attached incorrectly.

With the carboxylic acid on the outer shell of the gold nanoparticles, many biological molecules can be attached onto the nanoparticles in the second step of functionalization. Many peptides and proteins contain free amine groups thanks to the common amino acid lysine. Carboxylic acid groups also form amide bonds with free amine groups in the presence of EDC Carbodiimide crosslinker (sometimes with NHS, *N*-hydroxysuccinimide) (Sanz et al., 2022). This reaction is particularly suitable for biomolecules because the reaction takes place in an aqueous medium under room conditions. Therefore, there should not be any damage to fragile biomolecules such as enzymes or antibodies.

However, some of you might wonder what happens if a protein molecule has a free thiol group, or a cysteine, in its amino acid sequence. The answer is that such a protein molecule can be directly attached onto the gold nanoparticle surface. In reality, cysteine tends to be found in pairs naturally and formed cystine with a disulfide bridge (–S-S–). This is to aid the protein molecules in folding. Luckily these disulfide bridges are not particularly strong and, in the presence of a reducing agent, free cysteines form under a chemical equilibrium. Also, thanks to modern protein engineering, molecules with a specific amino acid sequence can be designed in laboratories. It is possible to "insert" a cysteine unit into the amino acid sequence of a protein molecule.

For other functionalities, we can use a generic method by using molecules of HS-R-X, where X is a reactive group including $-NH_2$ or –Br. Once these reactive groups are on the outer shell of the nanoparticles, standard organic chemistry can be used for multistep functionalization to provide the necessary functional groups. Since the gold core is quite stable and the protective inner shell should be able to prevent the nanoparticles from aggregation, many organic reactions can be carried out with these Au-HS-R-X nanoparticles.

1.6.2 Functionalization of Quantum Dots

Quantum dots are semiconductor-based nanoparticles. Since there are several different types of quantum dots, we shall focus on the most commonly used example, CdSe quantum dots coated with ZnS, or CdSe/ZnS, due to their emission spectrum on the visible range (Hines and Guyot-Sionnest, 1996). As a result, we need to investigate the chemistry of ZnS in order to attach a functional group onto the surface of this type of quantum dots.

Surprisingly, there is a certain degree of similarity to the functionalization of gold nanoparticles, owing to the sulfide nature of the ZnS shell. Thiol not only forms a strong bond with gold but also forms –S-S– disulfide bonds with the sulfur atoms on the surface of the ZnS shell. Therefore, many of the functionalization strategies for gold nanoparticles are applicable to CdSe/ZnS quantum dots. For example, thiolated carboxylic acid can also be used to functionalize CdSe/ZnS quantum dots (Sun et al., 2006).

In addition to thiols, several other functionalization strategies can also be used to functionalize CdSe/ZnS quantum dots. One trivial method is to exploit the interaction between phosphine or phosphine oxide and the ZnS shell. This interaction has been used during the preparation of quantum dots when the phosphines and phosphine oxides are used as the surfactant templates. If the other end of the surfactant has attached with a carboxylic acid group, similar nanoparticles to the quantum dots capped with thiolated carboxylic acid can be prepared. Waiskopf et al. had already demonstrated such a strategy (Waiskopf et al., 2011). Also, unlike thiolated carboxylic acids, phosphines have three alkyl chains and a maximum of three carboxylic acid groups can be attached to each phosphine molecule. Other functionalities on the phosphine molecules can also be added (Tomczak et al., 2009).

Also, thanks to the Zn component of the ZnS shell, use of strong chelating molecules can also be used to functionalize the CdSe/ZnS quantum dots. One classic example is engineered peptides with histidine units (Clapp et al., 2004). A histidine has an imidazole group and a strong chelating site to transition metals is created when multiple histidines are adjacent to each other (see Fig. 1.8).

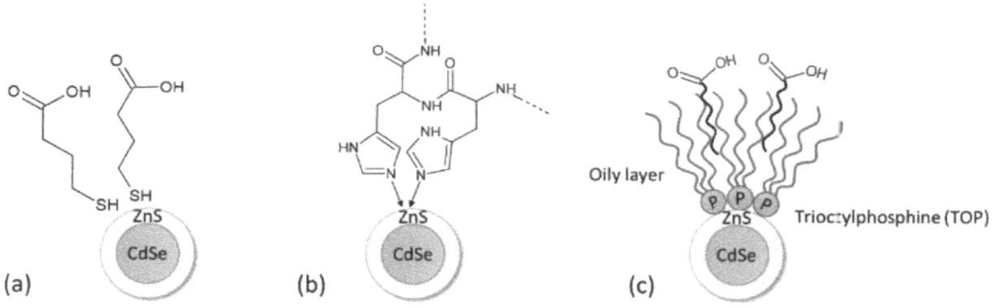

Figure 1.8 Three different strategies for functionalizing the quantum dot surface, using (a) thiolated carboxylic acid, (b) peptide with a dihistidine unit, (c) oleic acid bound onto the oily layer.

Indirect functionalization using "oily" functional molecule has also been reported. Since quantum dots can be prepared in a micelle template, formed by an "oily" layer of phosphine (TOP) and phosphine oxide (TOPO), and functional molecule with a long aliphatic tail could associate with this oily shell, leaving the polar functional group outside the shell. For example, a long-chain amine such as dodecylamine, or long-chain carboxylic acid (e.g., oleic acid), can be used for this purpose and the quantum dots will be covered with amine functional groups (see Fig. 1.8), which can then be used in many second step functionalization, including protein binding (Wuister et al., 2003).

1.6.3 Functionalization of Silica Nanoparticles

The functionalization of a silica surface is completely different from a gold or ZnS surface. On a silica surface, only Si–OH groups are available for activation. The use of thiol is not applicable here as there is no strong interaction between –SH and the Si–OH groups. There is no transition metal component and chelating groups are also of no use here. However, the functionalization of silica nanoparticles is far easier than many other nanoparticles thanks to the well-developed silane chemistry.

The most straightforward, also the most common, method to functionalize a silica surface is the use of alkoxysilane. A functionalized alkoxysilane usually has a Si center and three -OR groups attached directly onto the Si center. The remaining site on the Si is to provide various functionalities. Hence, a functionalized alkoxysilane has a general formula of $(RO)_3$-Si-X. Usually the alkoxyl groups are methoxyl or ethoxyl, and methanol or ethanol is produced as a byproduct during the reaction. Since there is a wide range of end groups (–X) to choose from, silanization is a versatile method and it is also applicable to other inorganic oxides, including iron oxides (see the next section). It also applies to various forms of silica, irrespective of whether they are nanoparticles or nanoporous solids. Details of the silanization method on silica will be discussed in Chapter 2.

1.6.4 Functionalization of Iron Oxide Magnetic Nanoparticles

There are several functionalization pathways for modifying the surface of iron oxide magnetic nanoparticles. Figure 1.9 summarizes several common functionalization methodologies for iron oxide magnetic nanoparticles. The choice of functionalization methodology is usually determined by the end use of the nanoparticles and the required functional groups on the surface.

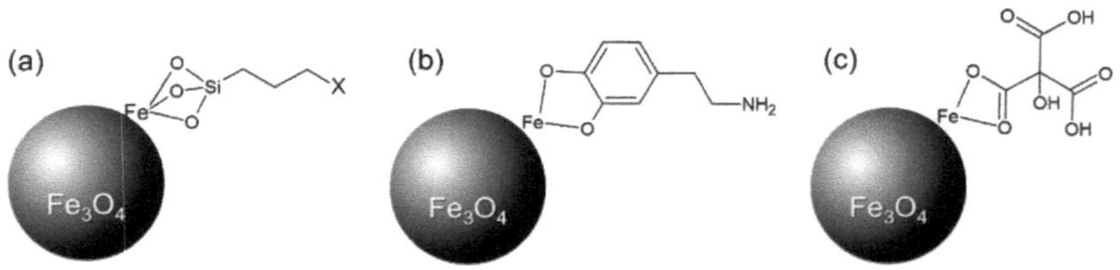

Figure 1.9 Illustration showing three surface functionalization methods for iron oxide (Fe_3O_4) nanoparticles using (a) organosilane, (b) dopamine, and (c) citric acid.

Among these methods, the use of natural polymer, in particular dextran, tends to be the most popular method. This is partly because dextran is a highly biocompatible material and relatively easy to handle. Dextran-coated iron oxide nanoparticles have been made commercially available recently. Endorem, 5–8 nm Fe_3O_4 nanoparticles coated with dextran, is a commercial contrast agent for MRI scans. Polysaccharides, including dextran, have a large amount of hydroxyl, –OH, groups on the polymer skeleton.

Therefore, it is logical to make use of these groups. Hydroxyl groups react with alkyl halides readily in the presence of heat following the below equation:

$$R\text{-}OH + R'\text{-}X \rightarrow R\text{-}R' + H_2O + HX \quad (1.4)$$

This property was particularly useful in substituting these hydroxyl groups with carboxylic acid groups, for ion exchanging or protein binding. For example, carboxymethyl cellulose (CMC) can be synthesized by using chloroacetic acid ($Cl\text{-}CH_2\text{-}COOH$) with cellulose and it is also commercially available. Similarly dextran, on iron oxide nanoparticles or as a free polymer, can be activated with the same pathway. Dextran-coated iron oxide nanoparticles have been converted into carboxylic acid-grafted nanoparticles using this method, as carboxymethyl dextran.

Some polysaccharides, such as chitosan, have other functional groups other than hydroxyl groups for activation. For example, chitosan has one $-NH_2$ group on every unit of the polysaccharide skeleton. Because of the difference in chemistry, this feature allows various functionalization pathways that are not possible for dextran. One common use of $-NH_2$ groups in bioscience is to graft fluorescent tags onto the polymer. Amine groups react readily with isothiocyanate (ITC) groups (-NCS) at room temperature. Common fluorescent tags such as fluorescein and rhodamine B are available in ITC forms, e.g., fluorescein isothiocyanate (FITC). The grafting protocol for tagging FITC onto proteins can be used for tagging chitosan (see Fig. 1.10). Another useful chemical property of amine groups is their reaction with aldehydes to form imine at room temperature. In the literature, there are plenty of examples of the use of various carbohydrates as a functional coating to functionalize iron oxide nanoparticles (Muthiah et al., 2013; Lachowicz et al., 2017).

There are also several ligands capable of chelating onto the iron atoms on the surface of the iron oxide nanoparticles. For example, carboxylic acid can coordinate onto the iron oxide surface (see Fig. 1.9c). Indeed, the strategy of using oleic acid to stabilize iron oxide nanoparticles is an example of the coordination chemistry of carboxylic acid onto iron oxide surface. One specific ligand that can be used for chelating iron oxide nanoparticles is dopamine. Similar to carboxylic acid, the catechol group on dopamine coordinates onto the iron atoms on surface (Fig. 1.9b). The amine groups on the other end of the dopamine molecule can be used for further functionalization. Likewise, molecules with a catechol group can also be used for functionalizing iron oxide surface. Dopamine has a particular high affinity towards iron so it is rather useful for functionalizing iron oxide nanoparticles.

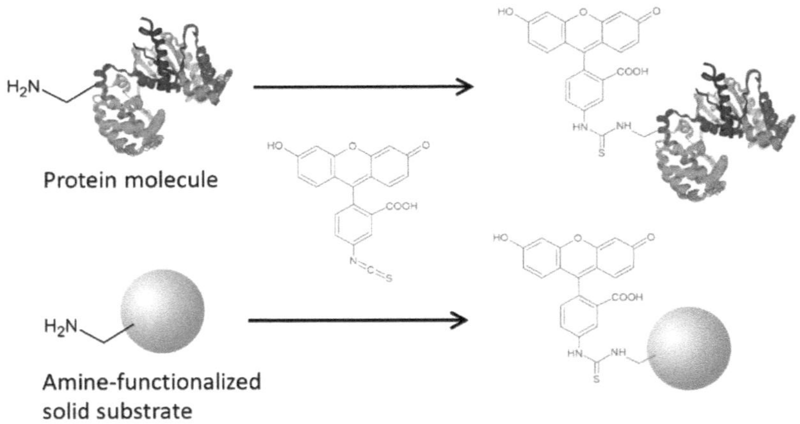

Figure 1.10 Schematic showing the binding chemistry of FITC onto a protein molecule and an amine-functionalized surface.

Using coordination chemistry avoids many undesired reaction conditions including high temperature. However, the stability of these coordinated iron oxide nanoparticles tends to be lower due to the lower bond strength than common covalent bonds. When there is a strong ligand, e.g., EDTA, in the system, such coordination bonds can be weaken and replaced by a stronger ligand. In this case, for example, a new Fe-EDTA bond may be formed by replacing Fe-dopamine. Therefore, choosing the right binding/functionalizing strategy is vital for the stability of the coating and the system.

The final and rather popular method is silanization (Fig. 1.9a) using organosilanes (Yiu, 2011). The advantages of silanization are that (1) covalent bonds are formed between the ligand and the iron oxide surface, making the functionalization more permanent;

(2) a wide variety of silanes are available so a broad spectrum of functionality can be grafted in a one-step reaction. Silanization is widely used in functionalizing silica surface and hence we have a detailed discussion in Chapter 2. The grafting chemistry and procedure are largely the same, no matter it is on iron oxide or silica surface.

1.7 Applications of Templated Nanoparticles

Because of the small size of these nanoparticles, they have tremendous commercial potentials in many areas. Since their applications can be very diverse, we highlight some key applications based on the chemical nature of these nanoparticles.

1.7.1 Polymer Latex Particles

Many polymer latex particles are not chemically active, except those that are designed to be. For example, polystyrene latex particles are rather inert and they are unlikely to break down in normal conditions. Because of this, in addition to their small size, polystyrene latex particles have been proposed to be used as a "lubricant" for large scale industries, including in oil industries as drilling fluid. However, in recent years, concern about plastic wastes in the environment, in particular microplastics, has been a global issue. These polystyrene particles are non-biodegradable and persistent in the environment for hundreds or even thousands of years. The use of a large quantity of non-biodegradable particles such as polystyrene with little control in disposal is likely to be heavily regulated very soon.

On the other hand, soft, biodegradable latex particles such as polycaprolactone (PCL), polylactic acid (PLLA) and polyglycolic acid (PLGA) are more likely to be used. These polymers are built from either ester or amide bonds between monomers. These ester and amide bonds can be "digested" or hydrolyzed by bacteria in the environment and also sunlight. Therefore, they are much less persistent than other polymers manufactured from additional polymerization such as polystyrene and polyethylene, and possess lower environmental burden.

Polycaprolactone had been used in agricultural research, delivering herbicides and pesticides to crops (Grillo et al., 2012). Capsules made from PCL allow these bioagents to be delivered to their target while conventional methods (e.g., spraying) lead to excessive use of these chemicals. This also enhances the efficiency of these bioagents (per kg) and the degraded products (caprolactone) are not considered as a pollutant to the environment (soil in this case). We can see the potential of using biodegradable latex particles in this area. However, the production cost of these particles is still an issue. A high concentration of degraded products (e.g., lactic acid) localized close to the plants could still be a problem, considering these plants are to be consumed by humans. More research is still needed for this application.

Biodegradable poly(lactic acid) PLLA, poly(glycolic acid) PLGA and their copolymers have been widely used in tissue engineering. This is not only because they are biodegradable but also because the metabolic products (lactic acid and glycolic acid) are not considered to be toxic to human body. Another advantage is that the degradation rate can be tuned by varying the lactic:glycolic acid ratio, allowing controlled drug release from copolymer particles. Such system has been applied to deliver anticancer drugs for chemotherapy and growth factors for tissue engineering. (Danhier et al., 2012; Acharya and Sahoo, 2011; Gentile et al., 2014). We should note that not all these nanoparticles for delivery were made with surfactant templating, although templating would help the homogeneity of the final product.

Away from biological applications, polymer latex particles have also been used as templates for making other nanoscaled materials themselves. This is partly due to their narrow size distribution and partly due to their ease to be removed from an inorganic matrix after the formation of the newly templated material. Removal of polymer latex template can be done by solvation by an organic solvent (or solvent extraction), by digestion (acid, base or even enzymatic) or simply by combustion. Details for this application will be discussed in Chapter 4.

1.7.2 Gold Nanoparticles

In the past, gold was considered to be catalytically inactive. However, in early 2000, it was found that gold nanoparticles are indeed active for a wide range of catalytic reactions if they are supported on a metal oxide surface. One breakthrough discovery was reported by Haruta and his co-workers (Torres Sanchez et al., 1997) where they found that Au nanoparticles supported on hematite (α-Fe_2O_3) was active in the oxidation of CO to CO_2. This is a valuable find since CO is not only poisonous to humans but also to catalysts even in low concentrations. On the other hand, CO_2 is much less harmful. Since then, supported gold nanoparticles were found to be active in many redox reactions (see Table 1.2).

Table 1.2 Selected examples of supported Au NPs in various catalytic reactions

Fields of applications	Reactants or reactions	Support materials
Indoor air quality control	Odor (commercialized), CO	Fe_2O_3
	Sick house gases	TiO_2
Pollutant abatement	Dioxin oxidation-decomposition	Fe_2O_3
	NO reduction	Al_2O_3
	N_2O decomposition	Co_3O_4
H_2/energy carrier	Water-gas shift	ZrO_2, CeO_2
	CO removal	Al_2O_3, Mn_2O_3, Fe_2O_3
	Fuel cell anode	Carbon black
Chemical process	Hydrochlorination	$AuCl_3$/activated carbon
	Hydrogenation of C=C, C=O	ZnO
	Liquid-phase selective oxidation	Activated carbon
	Propylene epoxidation	TiO_2(anatase) and Ti-SiO_2

We must note that these Au catalyst examples are all supported catalysts and not made from surfactant templating. Indeed, reports on free, unsupported gold catalysts are few and far between, partly because their activities are usually lower than the supported counterparts. There are some exceptions; for example, Comotti et al. compared unsupported Au NPs prepared using glucose as a structural directing agent, and compared the activity in glucose oxidation with an Au/C catalyst (Comotti et al., 2004). The initial activity of unsupported Au NPs was indeed marginally higher than Au/C in <100 s. However, when the reaction continued for a longer period, Au/C outperformed Au NPs by 3 folds in terms of conversion after 3 h. Research on surfactant-templated Au nanostructures as catalysts are even rarer, partly because scientists believe that this is unlikely to be successful. Also, the reactions chosen for testing these Au nanostructures are unconventional. For example, Sau et al., prepared Au NPs (10–150 nm) using Triton X-100 as the surfactant template (Sau et al., 2001). The catalysts were tested on the reduction of a dye Eosin using $NaBH_4$, and the results suggested that the smaller the particle size of Au was, the more active it was. That is perfectly reasonable. Hu et al. used oleylamine as the surfactant template to prepare Au nanowires and tested the catalytic activities on oxidation reaction of various benzylic compounds using O_2 at 1 atm (Hu et al., 2011). It was found that the solvent had a strong effect on the % conversion, with 1,4-dioxane as the best solvent with 81.7% conversion. A wide range of other benzylic compounds were also tested using the same reaction but most showed rather low conversion. In both examples, no comparison was made with supported catalysts.

Gold, similar to a few other noble metals (e.g., Pd, Pt), has a particular affinity to a few elements, notably sulfur, S. Hence, sulfur has become one notorious poison for these catalysts due to its strong association to the metal surface. Also, sulfur compounds including H_2S and SO_2 need to be removed from reaction feeds, including petroleum (due to their damage to catalytic converters). Interesting this property can also be exploited positively in other areas of science. For example, Au, in particular Au NPs, can be easily functionalized with sulfur compounds, such as thiolated organics. With this route, an inorganic-organic hybrid nanoparticle can be made.

As mentioned in Section 1.6.1, Au NPs can be functionalized directly without binding reagents using thiolated organic compound, such as cysteine, which is a commonly found amino acid with a thiol residual group. Indeed, it is responsible for the bad smell of rotten eggs (proteins), due to the sulfur component. Similar to other thiolated compounds, cysteine has an affinity towards Au surface. Early on, scientists had been exploiting such a phenomenon for functionalizing Au surface using peptides with cysteines on their structure (or sequence) (Ihs et al., 1991). The same theory would apply to Au NPs. For example, in early 2000, Tkachenko et al. tested the binding of several cysteine terminated peptides on Au NPs. (Tkachenko et al., 2003). This paved a promising path for peptide-functionalized Au NPs for the next decades and now there are thousands of papers in the literature reporting research on peptide-functionalized Au NPs. Applications of peptide-functionalized Au NPs include detection of organic molecules, diagnosis (Parveen et al., 2012), medical imaging (Singh et al., 2018), biomarkers for cells, and drug delivery (Torchilin, 2014).

1.7.3 Quantum Dots

Thanks to their superior luminescent property, quantum dots have huge potential in imaging, diagnosis and tracking applications. For example, Her-2 antibody tagged CdSe quantum dots have shown a high selectivity in marking breast cancer cells. Quantum dots have also shown applications in DNA detection and identification, acting as a DNA sensor. Generally, it works by attaching a single-strand DNA of interest on to the quantum dot surface. The "complimentary" strand of the DNA will strongly bind to the quantum dot, causing alteration on the fluorescence signal. Since the detection of fluorescence can be very sensitive, more sensitive than conventional UV and visible light, this method can enhance the DNA detection sensitivity to 1 fM (10^{-15} M) level for DNA and 10 fM for siRNA. This is a remarkable improvement on spectroscopic current methods (at around 10^{-9} M or nM) (Su et al., 2014). This detection method has already been proposed and tested for diagnosis for lung cancer and gastric cancer. Since the principle for DNA detection is similar for most DNA strands, application of quantum dots can be easily adapted to other DNA-linked purpose, including forensics.

Another area of application for quantum dots is immunosensing. Using antibodies, immunosensing is a highly sensitive and selective method. This method has been widely used in biomedicine and biotechnology. For example, immunosensing is currently used in the diagnosis of cancers and viral diseases, including HIV infections. The principle is to use the specific binding of antibodies for receptors on cancer cell surface or virus and then generating a signal, either optical or electrical. Similar to DNA, antibodies can be attached to quantum dots and their binding onto cancer cells and viruses can be detected using fluorescence. For instance, a current study suggests that using a quantum dot-based immunosensing system can detect prostate specific antigen at a concentration as low as pg/mL level. (Malekzad et al., 2017). Early cancer diagnosis is essential for patient's full recovery and also improving patients' quality of life. Enhancing the diagnosis technology can only be welcome by the medical community and the general public.

Despite such advancements in research for quantum dots, their toxicity is still a major obstacle to developing quantum dots as an *in vivo* diagnostic tool. Both Cd and Se are highly toxic and leaching of these elements into our body is likely to cause severe health hazards. Many other elements for quantum dots, including Te, In, and Pb, are also toxic. ZnS and ZnO are probably the only semiconducting materials which are safe enough for *in vivo* application and have shown quantum effect from their nanoparticles. However, the excitation/emission wavelength of both ZnS and ZnO is on the high energy end of the electromagnetic spectrum. This means that a light source (usually a laser) of high energy is needed for excitation. This is also not practical as such high power lasers are also very harmful. ZnS coated CdSe quantum dots have shown a lower toxicity than uncoated CdSe quantum dots, but their long-term effect is still questionable. Until their toxicity has been explicitly scrutinized, *in vivo* applications of quantum dots is unlikely to be of great success.

On the other hand, *ex vivo* diagnosis using functionalized quantum dots shows a much more promising future for these fantastic nanomaterials. Quantum dots-based testing systems have already shown shorter response time and brighter signal to conventional testing systems for DNA sensing for example (Raichlin et al., 2011). In modern society, a fast, accurate, and simple testing system is needed for the increasing demands in quick diagnosis of various diseases including cancers, heart diseases and genetic disorders. These systems may soon be available in the market and off the shelf kits will be so common that many families may do their own blood tests.

1.7.4 Iron Oxide-Based Magnetic Nanoparticles

The majority of the applications of iron oxide nanoparticles are to exploit their unique magnetic properties. One conventional application of magnetic materials is in recording, including magnetic tapes and hard drives. Therefore, when advances in magnetic nanoparticles became apparent, scientists were working on various methods to assemble these nanoscaled magnetic materials to build the ultrahigh-density magnetic recording devices (Sun et al., 2002). However, very quickly, flash drive technology became available and it is operated in a completely different branch of science. Now, flash drive technology dominates the commonly used data storage devices. Development in magnetic nanoparticles for recording has stalled. Scientists' focus turned to other areas.

1.7.4.1 Magnetic separation

Separation of molecules can be a complicated and laborious process and therefore exploiting the physical properties of a stationary phase can be very attractive to industries, particularly when the target species are of high values. This applies to many biomolecules. Many biomolecular products such as proteins and enzymes are produced via extraction from natural biosources, including cell/bacterial cultures, animals, and plant sources. Extraction from these sources is likely to be carried out from complex matrices and the target species may be of a very low concentration. Therefore, traditional separation methods using liquid-liquid phase extraction or using column separation may not always be suitable.

The use of magnetic nanoparticles allows separation in the presence of a magnetic field and it is particularly useful where other sophisticated technologies are not available (Wierucka and Biziuk, 2014). However, for this purpose, the magnetic nanoparticles themselves require surface functionalization, which should be tailored to specific target molecules, in order to enhance the selectivity (Hola et al., 2015). This is to assure the highest possible purity and avoid further purification steps, which will only increase the production cost and time. In general, composite materials (e.g., particles) made by coating/blending magnetic nanoparticles with a polymer matrix with further modification on the surface can be used. For example, chitosan/SPION (superparamagnetic iron oxide nanoparticles) composites carry a positively charged surface have been used for purifying hyaluronic acid (Yang and Lee, 2007) and glycopeptides (Fang et al., 2014) using electrostatic binding. However, electrostatic binding is of low specificity and targeting functional group can enhance specific binding. Kim et al., demonstrated using antibody-functionalized magnetic nanoparticles for specific binding of HIV-1 p24 Gag protein, as a detection device. (Kim et al., 2008). Furthermore, such antibody-functionalized nanoparticles were used to target specific receptors on cell surface, as a method to separate specific cells from a complex culture. Such principle has been demonstrated in the separation of cancer cells (Scarberry et al., 2008) and T-cells (Takahashi et al., 2009) from complex matrices.

One popular strategy is using aptamers for capturing specific molecules (Du et al., 2015). Aptamers are short single-strand DNA or RNA sequences that are capable to bind specific proteins and carbohydrates. Therefore, the technology can be easier transferable to a wide range of targets. For example, aptamer-functionalized iron oxide magnetic nanoparticles have been used to purify activated protein C (APC) (Hamedani et al., 2022). Aptamer is expensive but the target molecules tend to be of high costs too. If the magnetic composites can be recycled and reused for multiple times, operational costs become viable in a small scale. Another recent development is using molecular imprinting (MIPs) for highly selective binding for biomolecules. Use of magnetic MIPs (or MMIPs) for biomolecule separations and purifications will be discussed in length in Chapter 5.

1.7.4.2 Magnetic drug delivery

Similar to other nanoparticles, the surface of magnetic nanoparticles can be functionalized to carry drug molecules and deliver to cells upon internalization. Drug binding can be achieved via electrostatic interaction or chemical binding with cleavable bonds by enzymes (e.g., amide or ester linkers). For ease of controlled release, electrostatic binding is preferred. Moreover, the targeting molecules or antibodies can also be bound onto the magnetic nanoparticle surface for targeting specific tissues or a specific area (e.g., the tumor) of a tissue/organ.

One unique feature of magnetic nanoparticles is again their magnetic property. With an external magnetic field, these magnetic nanoparticles can be driven to the desired area(s) or tissues due for treatments. In this way, the drugs can be delivered to the target. This concept is particularly important to toxic drugs, including most being used for chemotherapies. For example, doxorubicin (DOX) can be used to treat breast cancer but with unpleasant side effects, including vomiting, upset stomach, loss of appetite, mild diarrhea, and skin rash. This can be caused by the drug being delivered to the healthy tissues/organs. Other drugs for chemotherapies can have more severe side effects and patients are therefore to have an unpleasant experience. Targeted delivery of drugs can reduce side effects because the drugs are less likely to reach healthy tissues/organs. Another important issue for consideration is cost. Anticancer drugs are expensive and many of them are not available to the public health services worldwide due to the cost issue. However, since a large portion of drugs does not reach the affected tissues/organs, or the tumors in case of cancer, improving the delivery can dramatically reduce the cost of use, making them economically viable for patients relying on public health care.

Delivery of DOX has been shown possible as a polyetheneimine(PEI)-DOX conjugate using PEG-coated magnetic nanoparticles (MNP). A hydrazine link is used between the PEI-DOX conjugates and the PEG-MNP carriers. This linkage is pH sensitive, cleavable upon changes in pH. PEI is used to enhance loading and uptake by rat glioma C6 cells. DOX can also be rapidly released when the pH reduces (under more acidic condition) (Kievit et al., 2011). Another anticancer drug, daunorubicin or DNR, was also tested for delivery using tetraheptylammonium (THA) coated magnetic nanoparticles. The THA coating enhanced interaction between magnetic nanoparticles and the K562 cells. K562 cells are considered as "resistant" to particle uptake. The THA coating helped particle-to-cell interaction and enhanced uptake compared with controls (Wang et al., 2007b). For designing the magnetic nanoparticle surface, we have to consider the interaction between particles and cells in order to maximize the uptake and hence delivery efficiency.

Although targeted delivery using magnetic nanoparticles was shown to be successful and workable in animal models (mainly mouse models), it is still far from clinical trials. The drug carriers (i.e., functionalized magnetic nanoparticles) are considered to be safe for use. However, the strong magnetic field required for driving the magnetic nanoparticles could be the main obstacle, due to safety issues. Another issue is that the magnetic field has to be quite accurate to drive the particles to the target through blood vessels. Moreover, superparamagnetic nanoparticles, which carry no permanent magnetic moment, should flow through blood vessels easily. When they are under a strong magnetic field, they carry a magnetic moment and can aggregate with each other. Inside a narrow blood vessel (capillary), this could cause blockage and the consequence can be serious. Without a magnetic field to direct these magnetic nanoparticles, they tend to end up in the liver and the spleen (Ashraf et al., 2019). Therefore, unless these two organs are the ultimate destinations, other targeting mechanisms will be required, including tagging targeting molecules such as folic acid and antibodies. All these issues need to be fully tackled before magnetic targeting on drug delivery can be put forward as chemotherapy for the general public.

It is worth pointing out that there are many types of inorganic magnetic nanoparticles, regardless of whether they were prepared using templates or not. However, most research in drug delivery was limited to iron oxide nanoparticles (Fe_3O_4 or γ-Fe_2O_3) due to safety issues. There are other materials including Co- and Ni-based materials possess similar magnetic properties, iron oxide provides an extra advantage of having low toxicity in general. Since some iron oxide nanoparticles (e.g., Endorem®) had already been approved by the FDA, products based on these iron oxide nanoparticles would be more likely to be approved for use in the future.

1.7.4.3 Magnetic gene delivery and magnetofection

A specific type of drug delivery is to deliver a designed piece of DNA (gene) or RNA to repair faulty gene found in patients. Using this technology to treat patients is commonly termed as "gene therapy." This is usually used for patients who have genetic defects and no other treatment can be applied for curing the diseases. In simple terms, the principle is to deliver this small piece (sequence) of gene to the cells and let the cells repair themselves. To deliver these gene sequences, some cell penetrating reagents (vectors) need to be used. Generally, we can classify the methods into "viral" and "non-viral." Viral gene therapies use an inactive virus to deliver the gene sequence to cells. Viruses are built to invade cells and therefore, they are highly effective for delivering gene sequences (Schaffer et al., 2008). However, we can easily link viral gene therapies to the risk of using a virus. Therefore scientists started to search for non-viral vectors for gene delivery.

The basic principle of non-viral gene delivery is to use a positively charged (cationic) reagent to bind to the negatively charged gene sequences (DNA or RNA) and form a conjugated complex, ready for delivery. One successful reagent is a lipid-based reagent that can bind gene sequences and form small liposomes for delivery into cells. This is similar to the micelles that we discussed earlier. In the market, Lipofectamine, which is a combination of two cationic lipids, is widely available for this purpose (Cardarelli et al., 2016). The efficiency of Lipofectamine is high, but it also causes a certain percentage of cell deaths (Lv et al., 2006). Also, Lipofectamine is rather expensive (currently at over US$ 100 for 0.1 mL of reagent). Scientists are keen to search for cheap alternatives.

In theory, solid particles with a positively charged surface can also bind gene sequences, DNA and RNA. However, the size of the particles has to be small enough to be internalized into cells. Despite this, the uptake of solid particles can be slow and hence the efficiency of delivery is always lower than liquid-phase reagents such as Lipofectamine. Gene delivery using magnetic nanoparticles, termed "magnetofection," has been reported in early 2000 (Scherer et al., 2002). In general, positively charged polymer such as polyethyleneimine (PEI) was coated onto the surface of magnetic nanoparticles, usually Fe_3O_4 or γ-Fe_2O_3 due to safety reasons, via either physical or chemical binding. Using magnetic nanoparticles has a number of advantages:

1. Magnetic nanoparticles allow easy separation and recovery after gene binding (see the previous section for magnetic separation).
2. Uptake efficiency can be improved by means of magnetic field.
3. Magnetic nanoparticles can be tracked using MRI post delivery.
4. Localized delivery can be carried out using a magnetic field.

For example, Adams et al. showed that using a static/dynamic magnetic field can enhance the uptake and transfection efficiency in a neural stem cell system (Adams et al., 2015). More importantly, the use of magnetofection is considered to be of low toxicity. One obstacle still to be solved is the lower efficiency than liquid-phase vector or viral vector. Now the use of magnetofection is moving towards other systems, such as agriculture.

1.7.4.4 Magnetic resonance imaging (MRI) and monitoring

The magnetic property of iron oxide nanoparticles gives a unique application in medical sciences. Iron oxide, particularly magnetite Fe_3O_4, nanoparticles have been developed and commercialized as an MRI contrast agent. There are two types of responses (T_1 longitudinal relaxation and T_2 transverse relaxation) that an MRI instrument can measure and these two responses happen simultaneously. For imaging, T_1 is usually preferred due to a higher resolution. For more details, see Box 1.3. Use of iron oxide nanoparticles as a contrast agent is mostly associated with T_2 imaging. They were not suitable for the higher resolution T_1 imaging due to their lower relaxivity (r_1) value and a high r_2/r_1 ratio. To generate a "bright" image, a high r_1 and a low r_2/r_1 ratio are required. With 5 unpaired electrons in the electronic configuration, Fe^{3+} ions increase the r_1 value. Fe_3O_4 magnetic nanoparticles also show a high r_2 value due to their intrinsic high magnetic moment. These two factors combine, commercial Fe_3O_4 nanoparticles are unlikely for use as T_1 contrast agents. Theoretically, this can be made possible by reducing the particle size (from 20 nm of commercial particles for example). The magnetic moment at saturation for magnetic nanoparticles is affected by their crystallinity and crystal domain size. Reduction in domain size can reduce the volume magnetic anisotropy and spin disorders on the surface of the nanoparticles, leading to a lower magnetic moment at saturation. This critically reduced their r_2 value, making them unsuitable for T_2 contrasting due to low T_2 response. However, the r_2/r_1 ratio will therefore increase. Research reports suggest that Fe_3O_4 nanoparticles showed T_1 response with a particle size ranging from 1 to 12 nm; optimal T_1 response shown at 1–5 nm. At this size range, the nanoparticles' magnetic moment is effectively too low to distort the surrounding magnetic field (Bhavesh et al., 2015). If a Fe_3O_4 nanoparticles sample has a narrow size distribution (or good size control) of particle size within the 1–5 nm range, T_1 contrasting is possible, and that has been demonstrated in the literature (Kim et al., 2011).

Since T_1 imaging is of higher resolution, the use of T_1 contrast agents is popular for MRI diagnosis. In the commercial market, the most common T_1 contrast agents are gadolinium-based complexes such as gadopentate dimeglumine, gadoterate meglumine and gadoversetamide (Xiao et al., 2016). These complexes are gadolinium ion (Gd^{3+}) chelated with diethylenetriaminepentaacetate (DTPA), diethylenetriaminepenta-acetate bismethylamide (DTPA-BMA), and 10-(2-hydroxypropyl)-1,4,7,10-tetra-azacyclododecane-1,4,7-triacetate (HP-DO3A), respectively. These contrast agents are widely regarded as the safest options, all approved by the FDA in USA (U.S Food & Drug Administration, 2016; Dale et al., 2015). The unusually long electronic spin relaxation time and high magnetic moment of Gd^{3+} result in efficient relaxation of local protons, resulting in a highly effective contrast agent. A patient only requires a small dose of Gd complexes (around 0.15 mmol/kg of weight of patient) to produce a sufficiently high T_1 response during an MRI scan. However, the free or unchelated Gd^{3+} ion is toxic even at a relatively low dosage. Therefore, Gd^{3+} ions require binding to strong ligands. These ligands not only protect patients from the toxic free Gd^{3+} ions but can also influence the distribution of Gd^{3+} within the body by ligand design and alter the efficiency of Gd^{3+} in terms of shortening the T_1 and T_2 responses (De León-Rodríguez et al., 2016).

Box 1.3 Basic principles of MRI

Fundamentals of MRI

Magnetic resonance imaging (MRI) is widely used as a non-invasive medical image technique for accurate diagnosis of various diseases, including multiple sclerosis, defects in the blood-brain barrier and various cancers. MRI provides good image contrast, excellent spatial resolution and a prolonged observation time (Gómez-Vallejo et al., 2018). MRI can be viewed as a giant NMR spectrometer and the patient is your sample for NMR. Generally, an MRI system measures water proton relaxation under a strong magnetic field (0.2–2.0 Tesla), which is still much weaker than an NMR spectrometer (2.0–7.0 Tesla). Since human tissues are comprised mainly of water, protein and fat, and they contain suitable hydrogen atoms in molecules, allowing measurement on proton relaxation. When there is no external magnetic field applied, water proton dipole moments in a tissue (or sample) are orientated randomly. Once the sample is under a static main magnetic field (B_o), the magnetic dipole moments of protons are either pointed in the direction of the magnetic field (low-energy state) or against the direction of the magnetic field (higher energy state). The population of each state (high energy and low energy) of protons is dictated by the Boltzmann distribution, with a higher population in the lower energy state. The gap between the two energy states is directly proportional to the magnitude of the magnetic field applied. The difference in distribution across the two energy levels leads to a net magnetization (M_o).

When the low-energy protons are bombarded with radiofrequency (RF) energy perpendicular to the main magnetic field, causing the protons to shift to the higher energy state. At resonant frequencies of the RF energy, known as the Larmor frequency $\omega = B_0\gamma$, proton shifts to the higher energy state occurs. The gyrometric constant (γ) for the molecule being energized depends on the (bio)chemical nature of the material and the surrounding environment (i.e., the tissue), known as chemical shift. Hence, by bombarding a range of frequencies of RF energy at the water protons, a spectrum of the sample composition can be obtained. After the application of the RF energy, the excited nuclei are "stimulated" by the surrounding nuclei and return to their original energy states (lower energy). To return to the previous energy state, the proton releases an RF energy dependent on the gyrometric constant. This also affects the time taken for the proton to return to the lower energy state. Since the frequency of the emitted RF energy is the same as the bombarding protons, a "picture" or "map" can be generated from the radio waves emitted due to each frequency of RF energy representing a particular type of biological tissue (Dale et al., 2015).

T_1 and T_2 responses

The longitudinal T_1 response time (or spin-lattice relaxation time) refers to the time required for the system to return to 63% of its original magnetization, M_o, value after it has been subjected to an RF energy pulse. The transverse relaxation time, T_2 (or spin-spin relaxation) is defined as the length of time for the transverse component of M_o to decay to 37% of its original value (see Fig. 1.11). Generally, T_1 responses are dependent on the applied magnetic field strength B_o, while T_2 responses are not affected by B_o. T_2 relaxation is dependent predominantly on the magnetic dipole-dipole interactions and the diffusion constant of the protons. T_1 and T_2 relaxation occur simultaneously but independently but T_2 response time is usually much smaller than T_1 and T_2 decay usually outpaces T_1 recovery (Roth, 2012), as shown in Fig. 1.11. Because T_1 images usually have a greater resolution than T_2 images, T_1 imaging is often preferred.

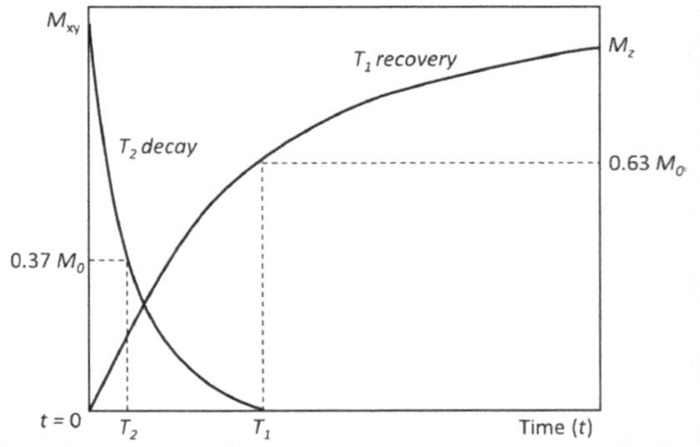

Figure 1.11 The relation between T_1 recovery and T_2 decay in MRI.

These Gd-based complexes generally have short circulation times and are quickly excreted from patient's body via the urinary tract (Kim et al., 2011). Hence, a relatively short window allows MRI imaging with these contrast agents. Even with strong ligand binding, Gd-based contrast agents are still not suitable for patients with liver or kidney conditions as they are usually excreted via the renal system a few days after injection or by glomerular filtration. Accumulation of toxic Gd^{3+} species, despite being chelated, cause the development of nephrogenic systemic fibrosis. Dialysis might be required to remove these Gd^{3+} species from patients, increasing the complexity and risks for patients. Therefore, non-Gd-based T_1 contrast agents, such as small Fe_3O_4 nanoparticles, are still sought after (Vergauwen, 2018).

Although less desirable in general, MRI diagnoses with T_2 response are still being carried out. Due to distortion on the magnetic field caused by the contrast agents, T_2 contrast agents are associated with darker images. Superparamagnetic Fe_3O_4 nanoparticles of size less than 50 nm can be used T_2 contrast agents. For example, Feridex and Endorem are essentially polymer (dextran) coated Fe_3O_4 nanoparticles. These T_2 contrast agents are administered slowly over a period of 30 min, with optimal imaging window from 30 min to 4 h post infusion, at a typical dosage of 5 mg/kg (Hubert et al., 2019). These Fe_3O_4 nanoparticles can be cleared from the blood by phagocytosis accomplished by the reticuloendothelial system (RES), resulting in uptake and localization in lymph nodes, spleen, bone marrow and liver. Since most lesions and internal scars do not have Kupffer's cells, or the ability to take up particles (Morana et al., 2007), Fe_3O_4 nanoparticles are unable to distort the magnetic field in this area, leading to an increase in T_2 signal. The image generated will be bright in the defected areas (Dale et al., 1995).

1.7.4.5 Hyperthermia as potential cancer treatments

One unusual property of magnetic nanoparticles is that they can generate heat under an alternating magnetic field. This is due to the change in the direction of the magnetic moment of nanoparticles when the external magnetic field is changing direction (Fig. 1.12). Scientists start to study and exploit this unique phenomenon for medical applications. Although the amount of heat energy generated from hyperthermia is not high (usually a few degrees Celsius), it could be enough to damage cancer cells and, ideally, cause cell death. Plenty of research work focuses on fine-tuning this process for cancer therapy, as summarized in Table 1.3.

Table 1.3 Examples of functionalized MNPs for hyperthermia (Thoidingjam and Tiku, 2017)

Magnetic nanoparticles	Cell line/ model	Observations/ findings	References
Starch coated MNPs	Epithelial cells (CRL-1666)	Effective ablation of metastatic tumors in the spine	Zadnik et al., 2014
Carboxymethyl dextran-coated MNPs	Breast cancer cells (MCF-7)	More effective than "hot water" in killing cancer cells	Rodríguez-Luccioni et al., 2011
Uncoated ferromagnetic and super-paramagnetic MNPs	Breast cancer cells (MCF-7)	Ferromagnetic MNPs showed higher efficiency than superparamagnetic MNPs in killing cells	Baba et al., 2012
Uncoated MNPs (Micromod)	*In vivo* model, mice with MTGB tumors	Slower tumor regrowth and less tissue damage compared with microwave hyperthermia	Petryk et al., 2013
Dextran-coated MNPs	*In vivo* model, mice with MTGB tumors	More effective than microwave hyperthermia in suppressing tumor growth. Combined treatment of hyperthermia with radiation was most effective	Giustini et al., 2011

Hyperthermia as a cancer therapy seems too good to be true. At the moment, there are still technical issues that need to be overcome before clinical trials. For instance, the amount of magnetic nanoparticles required to generate such heat, enough to cause cell death on tumors, is still high. Such a high level of injected magnetic nanoparticles may cause complication on patients' bodies, e.g., overloading the liver. Also, the heat generated might also cause damage to healthy tissues. This is a matter of targeting too. Currently, uncoated magnetic nanoparticles will still localize to the liver. Targeting reagents (see previous section) may help but that will only increase the cost of the therapy to an unrealistic level. Moreover, the magnetic field strength required for hyperthermia is very high. Such a high field strength may also cause damage to the patient's body, in particular the brain. Overall, hyperthermia as a cancer therapy still needs some significant developments before it is ready for clinical trial.

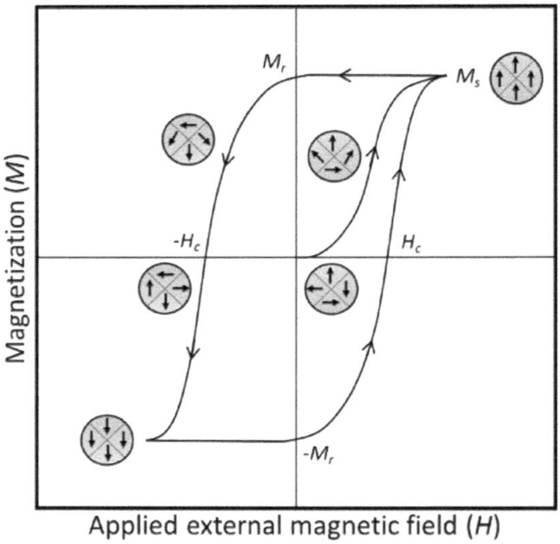

Figure 1.12 A typical M vs H diagram to show the changes in direction of magnetic moment when the external field is changing in magnitude and direction. H_c: Coercivity, M_s: Magnetization at saturation, M_r: Magnetic remanence.

Based on the same principle, another possible development is thermally controlled drug release. The principle is to coat a thermally sensitive polymer onto a magnetic nanoparticle. The polymer coating can absorb drug molecules and collapse at reduced temperatures. When the temperature increases, the polymer coating will open up and the drug will be released. The heat required for such temperature rise can be provided by the hyperthermia property of magnetic nanoparticle cores under alternating magnetic field. There are thermally sensitive polymers that can be suitable coating candidates. A classic example is poly(NIPAM), or poly(*N*-isopropylacrylamide), and various derivatives of poly(NIPAM) provide various ranges of physical (e.g., transition temperatures) and chemical properties (e.g., hydrophobilicity/hydrophilicity). Figure 1.13 depicts how such thermally sensitive drug delivery system works.

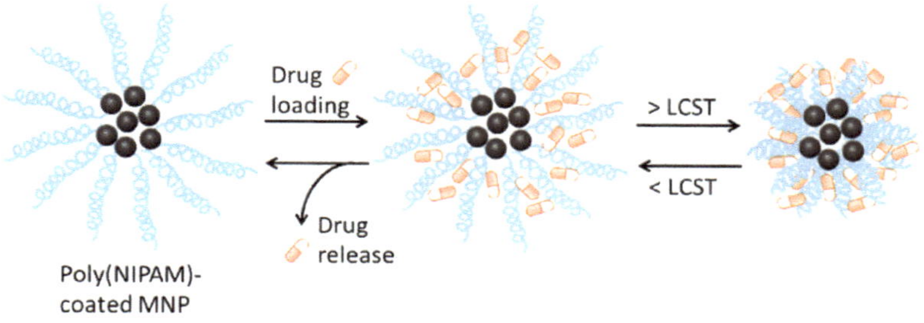

Figure 1.13 Illustration to demonstrate using poly(NIPAM) coated MNPs for temperature-stimulated drug release. LCST is the lowest critical solution temperature. For poly(NIPAM), LCST = 32°C.

In contrast to hyperthermia therapy discussed previously, the amount of magnetic nanoparticle required this thermal release therapy is much less. Practically, this is more likely to be applicable to a patient. Nonetheless, there are still technical issues with such systems. For example, although the "critical" temperature of the polymer coating for drug release can be tuned, it was usually done via a co-polymerization with another polymer strand. This not only increases the complexity of the coating but, in some cases, causes an increase in toxicity. Another issue is still the risk associated with the strong magnetic field that is to be used for hyperthermia, as mentioned previously.

1.7.5 Alloy-Based Magnetic Nanoparticles

Recent advancements of magnetic nanoparticles, in terms of both complexity and production, have been driven by commercial demand. Magnetic properties of materials have been exploited for data storage and memory (recording tapes, hard drives) for decades, but with the demand for the ever-reducing particle size in order to build more compact devices with a higher storage capacity, magnetic nanoparticles have become highly sought after. Back in the mid-1950s, when IBM built the first magnetic hard drive, the total storage capacity is 5 MB with a density of 2 kbit/in^2 (or ca. 40 B/cm^2). This is unimaginable for today's flash drive capacity. With such advancements in nanotechnology, scientists are aiming to reach a storage density of a massive Tbit/in^2 level. To achieve this, magnetic nanoparticles of diameter < 4 nm have to be used for coating the hard drive surface.

Iron oxide nanoparticles are unlikely to fit the bill because of their relatively low magnetic anisotropy. Therefore, magnetic alloy nanoparticles such as FePt and CoPt are to be considered as possible candidates. Take FePt nanoparticle as an example, the face-centered tetragonal (fct) phase has a high magnetic anisotropy value ($K = 4\text{–}10 \times 10^7$ erg/cm^3). And more importantly, FePt is chemically stable at room conditions. Degradation of the magnetic properties is unlikely to be an issue. However, as seen in using iron oxide nanoparticles for magnetic recording devices, new technology using non-magnetic materials become more popular and available. Development in this area did not go as fast as other areas. Research on applications of FePt nanoparticles has been on biomedicine too such as hyperthermia (Roca et al., 2009), similar to iron oxide nanoparticles. It also has shown applications in catalysis (Mazumder et al., 2010).

1.8 Templated or Template Free?

An increasing number of protocols allow template-free synthesis of various kinds of nanoparticles. Templated and template-free syntheses have their pros and cons, which are summarized in Table 1.4. The choice of protocol largely depends on the further modification requirements and the end applications. Generally, if a precise size and size distribution are required for the end application, a templated synthesis will be recommended. If the end applications are dictated by the surface functionalization, and the precision of particle size is not a critical factor, template-free synthesis will cause fewer problems downstream (e.g., functionalization/modification, template removal for toxicity reasons or storage).

Table 1.4 Comparison between templated and template-free synthesis of nanoparticles

Templated	Template free
Precise particle size	Particle size can be difficult to control
Narrow particle size distribution	Particle size distribution can be wide
Removal of unwanted surfactant/template is required	No template to remove
Surface functionalization can be hampered by template	Straightforward surface functionalization
Template toxicity	No concern in template toxicity

1.9 How Safe Are Nanoparticles?

Many people ask, "how safe are nanoparticles?" This is a tricky question because of the complexity of nanoparticles. There are many reports published recently concerning the safety and the toxicity of nanoparticles. The results are mixed and sometimes contradictory; some found nanoparticles to be toxic while others found them safe to use even *in vivo*. This is not surprising to many scientists in this field because the nanoparticles are usually tested for only one particular property or cell response. Naturally, the toxicity of nanoparticles varies from one type to another. In fact, there are many factors influencing the toxicity of nanoparticles. Some factors are trivial but the others can be less obvious. We shall take a few paragraphs to discuss this important issue.

First, we need to understand that the word "nanoparticles" may mean many different types of materials. To say nanoparticles are harmful to human is similar to saying "all chemicals are dangerous," which is obviously not entirely true. For example, gold nanoparticles are clearly less toxic than TeSe-type quantum dots because gold is a material of lower toxicity than TeSe, which is highly toxic as a bulk material. As a result, gold nanoparticles have been one of the most used nanoparticles for *in vivo* studies. Also, it is unfair to say that nanoparticles are toxic simply because quantum dots are toxic. Therefore, the toxicity of nanoparticles is highly dependent on their core material. Again, take quantum dots as examples, ZnS coated CdSe quantum dots are less toxic than uncoated CdSe quantum dots. This is because ZnS is much less toxic CdSe itself. If the ZnS shell can fully cover the CdSe core, and be stable in physiological conditions, the toxicity of the quantum dots is much lower. More importantly, although ZnS is another semiconducting material, the ZnS coating does not interfere with the CdSe emission property. Other nanoparticles also adapt this design with a "protective coating to reduce their toxicity. This coating can be silica or organic polymers, e.g., dextran. Since the chemistry or chemical properties of the nanoparticle cores vary so widely, we must understand the materials chemistry of each nanoparticle species before making judgment.

Next is the size factor. It is generally believed that the smaller the size of a nanoparticle, the more toxic the nanoparticle is. This is because nanoparticles of size smaller than 150 nm can penetrate through the cell membranes and enter the cytoplasm. This phenomenon is called internalization or uptake. The nanoparticles in cytoplasm may interfere with some chemical reactions inside a cell, e.g., producing "reactive oxygen species" (ROS), and hence induce toxicity. Moreover, the rate of internalization usually increases as the particle size decreases, so smaller nanoparticles are believed to be more toxic than larger ones. Quite often, nanoparticles for biological use are coated with some "biocompatible" materials to reduce their toxicity. Natural polymers, including polysaccharides and peptides, are commonly used as coating materials for this purpose. Among synthetic polymers, polyethyleneglycol or PEG is one of the most popular coating materials for biocompatible nanoparticles. This leads to the next factor, the functionality on surface. Nanoparticles of the same core material but different surface groups can have a different degrees of toxicity. For example, silica nanoparticles were found to have reduced toxicity when the surface was modified with APTES (or amine-functionalized SiO_2). This is because of the positive charge generated from the amine groups. (Morris et al., 2016). However, research also found that the toxicity of SiO_2 was not changed due to the amine functional groups (Yu et al., 2012). Again, the picture is still unclear.

Moreover, the end use of the materials also affects our definition of the word "safety." For example, using the nanoparticles for drug delivery requires a system that has to be proved safe to human body while nanoparticles for DNA analysis is an *ex vivo* application. It may be advisable to test the toxicity of the designed nanoparticles in the particular system for the end use. This means that nanoparticles for breast cancer diagnosis need to be tested in breast cells or breast tissue prior for use. Such tests are not straightforward because the toxicity of all functionalized nanoparticles is different to different targets (i.e., different tissues in the human body). Therefore, it can easily confuse users that one sample safe to a particular tissue may be harmful to another part of the body, or a slight change on the functionality or even size of shape of the particles may significantly increase the toxicity.

It is also important to understand that the toxicity of most materials is dose-dependent; this includes drugs. This means that under a certain dose limit, the materials will have a limited harmful effect on humans. Nanoparticles are not exceptions to this rule. LD50 (lethal dose to kill 50% of the cell population) is a good metric for such measurement. For example, arsenic, a known toxic material, has an LD50 about 15 mg/kg on rats while that for sodium chloride (table salt) is around 3 g/kg. One can vaguely interpret these numbers as arsenic is roughly 200 times more toxic than sodium chloride. That does not mean that sodium chloride is entirely safe for consumption. For nanoparticles, LD50 values vary greatly, due to the wide variety of nanoparticles. For instance, chitosan-coated Au nanoparticles showed a value of 2 g/kg in one research paper (Pokharkar, et al., 2009), and it seems very benign. On the other hand, silver nanoparticles (Ag NPs) were found to have an LD50 as high as 5 g/kg via oral administration (Maneewattanapinyo, et al., 2011), even though most researchers considered Ag NPs to be toxic. Based on these LD50 data, Ag NPs are only just as toxic as sodium chloride, which is common table salt to the public. Let us consider this, it is very likely that we will take 1 g of salt in meals, every day (the current recommended dose is 1.5 g per day) (American Heart Association, 2020). However, taking a dose of 100 mg of Ag NPs a day is very unlikely, and certainly not recommended. So, how likely will we be poisoned by nanoparticles?

One final problem is the contaminants in a nanoparticle sample subjecting to toxicity assay. In order to make the nanoparticles "look" more stable, many commercial nanoparticle samples/

products were stabilized by additives, such as surfactants and other stabilizing reagents, in solutions. The identity of these reagents is unlikely to be disclosed by the manufacturers (or suppliers) due to the commercial value of their product. This is similar to listing all food additives in packaging, many consumers have little idea *exactly* what they have purchased. However, in some cases, these additional reagents are more toxic than the nanoparticles themselves, causing negative results in toxicity tests. Since the identity of these reagents is not known, misinterpreted results due to these additives become very possible.

Measuring toxicity is a complex process. Many published research works on nanoparticles include a "cytotoxicity" section, which means the dose is required to trigger significant cell deaths. In general, these are carried out using commercial assay kits, e.g., MTT, WST-1, prestoblue, calcine AM, and there are many available. All these kits are based on various cell behaviors and they target different cell activities. However, the majority of these are color based, which means that the life or death of a cell will trigger some color change in the assay. The color change is commonly measured by a spectrophotometer. For soluble, transparent materials, these assay kits are useful tools for a quick assay on toxicity. They are easy to use and widely used in science communities. However, for insoluble solid materials, these kits become inaccurate because the test materials themselves could scatter or adsorb some incident light. It would be worse if the test materials also adsorb a certain amount of the colored signal molecules. This problem is particularly apparent for black materials such as Fe_3O_4 magnetite nanoparticles and carbon nanomaterials. Therefore, contradictory results in toxicity are commonly found in this area. One recent research highlighted the problems with color assays when Podgórski et al. used 5 different commonly used assays to assess the viability of cells grown on three widely used polymer materials for tissue engineering: poly(lactic acid) PLLA, polyurethane PU and polycaprolactone PCL (Podgórski et al., 2022). When the cell viability was tested after 24 h, viability of as low as 30% was recorded (a benchmark of 70% was considered to be viable) but these cell cultures were healthy and continued to proliferate well after 7 days. This highlights the inaccuracy of colorimetric assays for cytotoxicity. More accurate measurements are found using *in vivo* models such as zebrafish (*Danio rerio*) where many nanoparticle species were found to have an LD50 at around 10–50 mg/L, which is considered to be relatively safe.

1.10 Future Outlook

The wide variety of nanoparticles, irrespective of whether they were made with or without templates, have shown a broad spectrum of physical and chemical properties. With robust functionalization protocols available, designing the "desired" nanoparticles for a specific task is now become very likely, although still challenging. The small size of nanoparticles also means that they can perform certain tasks (e.g., drug delivery) that other materials cannot. We would expect various nanoparticles will become commonly used materials in many daily life applications in the very near future. There are still several obstacles that slow down the development of nanoparticles.

Compared with bulk materials, nanoparticles are still relatively expensive. This is because the production and functionalization of nanoparticles are still on small scales. It is partly due to the precision required for functionalized nanoparticles. Also, some components in these functionalized nanoparticles are expensive, e.g., gold core and tagged antibody. Its production and purification are also labor-intensive. Therefore, production cost and scaling are the first obstacles to be overcome.

But more difficult to overcome is the skepticism from certain sectors in the public. As seen in the previous section, safety of nanoparticles is still a concern despite published reports about their low toxicity. Perhaps there are risks associated to the use of nanoparticles, but most cytotoxic drugs including those used in chemotherapies are also posing certain risks. Other therapies, including radiotherapies and electrotherapies, also carry some risks. The public needs to understand and accept that all new therapies will be associated with some risks. Whether that such risks can outweigh the damage that caused by the diseases need to be considered by the patients, their family, and their doctors.

For some cancer patients, the consequence without a therapy could be death or long-term suffering. For other cases, the consequence may not be so bad. Therefore, consideration has to be made on a case-by-case basis.

References

Acharya, S., Sahoo, S. K. (2011). PLGA nanoparticles containing various anticancer agents and tumour delivery by EPR effect, *Adv. Drug Deliv. Rev.*, **63**, pp. 170–183.

Adams, C. F., Rai, A., Sneddon, G., Yiu, H. H. P., Polyak, B., Chari, D. M. (2015). Increasing magnetite contents of polymeric magnetic particles dramatically improves labeling of neural stem cell transplant populations, *Nanomedicine: Nanotechnol. Biol. Med.*, **11**, pp. 19–29.

American Heart Association 2021. Reducing Sodium in the U.S. Diet. https://www.heart.org/-/media/Files/About-Us/Policy-Research/Fact-Sheets/Access-to-Healthy-Food/Reducing-Sodium-in-the-US-Diet-Fact-Sheet.pdf (accessed 30 August 2022).

Ashraf, S., Taylor, A., Sharkey, J., Barrow, M., Murray, P., Wilm, B., Poptani, H., Rosseinsky, M. J., Adams, D. J., Lévy, R. (2019). *In vivo* fate of free and encapsulated iron oxide nanoparticles after injection of labelled stem cells, *Nanoscale Adv.*, **1**, pp. 367–377.

Autumn, K. Gravish, N. (2008). Gecko adhesion: evolutionary nanotechnology. *Phil. Trans. R. Soc. A*, **366**, pp. 1575–1590.

Baba, D., Seiko, Y., Nakanishi, T., Zhang, H., Arakaki, A., Matsunaga, T., Osaka, T. (2012). Effect of magnetite nanoparticles on living rate of MCF-7 human breast cancer cells, *Colloids Surf. B*, **95**, pp. 254–257.

Badia, A., Singh, S., Demers, L., Cuccia, L., Brown, G. R., Lennox, R. B. (1996). Self-assembled monolayers on gold nanoparticles, *Chem. Eur. J.*, **2**, pp. 359–363.

Bhavesh, R., Lechuga-Vieco, A. V., Ruiz-Cabello, J., Herranz, F. (2015). T_1-MRI fluorescent iron oxide nanoparticles by microwave assisted synthesis, *Nanomaterials*, **5**, pp. 1880–1890.

Cardarelli, F., Digiacomo, L., Marchini, C., Amici, A., Salomone, F., Fiume, G., Rossetta, A., Gratton, E., Pozzi, D., Caracciolo, G. (2016). The intracellular trafficking mechanism of Lipofectamine-based transfection reagents and its implication for gene delivery, *Sci. Rep.*, **6**, 25879.

Clapp, A. R., Medintz, I. L., Mauro, J. M., Fisher, B. R., Bawendi, M. G., Mattoussi, H. (2004). Fluorescence resonance energy transfer between quantum dot donors and dye-labeled protein acceptors, *J. Am. Chem. Soc.,* **126**, pp. 301–310.

Comotti, M., Della Pina, C., Matarrese, R., Rossi, M. (2004). The catalytic activity of "naked" gold particles, *Angew. Chem. Int. Ed.,* **43**, pp. 5812–5815.

Dale, B. M., Brown, M. A., Semelka, R. C. (2015). *MRI: Basic Principles and Applications*, 5th ed. (John Wiley & Sons, Ltd., UK).

Danhier, F., Ansorena, E., Silva, J. M., Coco, R., Le Breton, A., Préat, V. (2012). PLGA-based nanoparticles: an overview of biomedical applications, *J. Control. Release,* **161**, pp. 505–522.

De León-Rodríguez, L. M., Martins, A. F., Pinho, M. C., Rofsky, N. M., Sherry, A. D. (2015). Basic MR relaxation mechanisms and contrast agent design. *J. Magn. Reson. Imaging,* **42**, pp. 545–565.

Du, F., Guo, L., Qin, Q., Zheng, X., Ruan, G., Li, J., Li, G. (2015). Recent advances in aptamer-functionalized materials in sample preparation, *TrAC - Trend Anal. Chem.,* **67**, pp. 134–146.

Duan, R., Wang, B., Zhang, T., Zhang, Z., Xu, S., Chen, Z., Lou, X., Xia, F. (2014). Sensitive and bidirectional detection of urine telomerase based on the four detection-color states of difunctional gold nanoparticle probe, *Anal. Chem.*, **86**, pp. 9781–9785.

Fang, C., Xiong, Z., Qin, H., Huang, G., Liu, J., Ye, M., Feng, S., Zou, H. (2014). One-pot synthesis of magnetic colloidal nanocrystal clusters coated with chitosan for selective enrichment of glycopeptides, *Anal. Chim. Acta,* **841**, pp. 99–105.

Gentile, P., Chiono, V., Carmagnola, I., Hatton, P. V. (2014). An overview of poly(lactic-co-glycolic) Acid (PLGA)-based biomaterials for bone tissue engineering, *Int. J. Mol. Sci.*, **15**, pp. 3640–3659.

Giustini, A. J., Petryk, A. A., Hoopes, P. J. (2011). *In vivo* biodistribution of iron oxide nanoparticles: an overview, *Prog. Biomed. Opt. Imaging - Proc. SPIE*, **7901**, p. 790117.

Gómez-Vallejo, V., Puigivila, M., Plaza-García, S., Szczupak, B., Piñol, R., Murillo, J. L., Sorribas, V., Lou, G., Veintemillas, S., Ramos-Cabrer, P., Llop, J., Millán, A. (2018). PEG-copolymer-coated iron oxide nanoparticles that avoid the reticuloendothelial system and act as kidney MRI contrast agents, *Nanoscale,* **10**, pp. 14153–14164.

Grillo, R., dos Santos, N. Z. P., Maruyama, C. R., Rosa, A. H., de Lima, R., Fraceto, L. F. (2012). Poly(eopen-caprolactone)nanocapsules as carrier systems for herbicides: physico-chemical characterization and genotoxicity evaluation, *J. Hazardous Mater.*, **231–232**, pp. 1–9.

Hamedani, N. S., Happich, F. L., Klein, E.-M., Rühl, H., Mayer, G., Oldenburg, J., Müller, J., Pötzsch, B. (2022). Aptamer loaded superparamagnetic beads for selective capturing and gentle release of activated protein C, *Sci. Rep.*, **12**, p. 7091.

Hines, M. A., Guyot-Sionnest, P. (1996). Synthesis and characterization of strongly luminescing ZnS-capped CdSe nanocrystals, *J. Phys. Chem.*, **100**, pp. 468–471.

Hola, K., Markova, Z., Zoppellaro, G., Tucek, J., Zboril, R. (2015). Tailored functionalization of iron oxide nanoparticles for MRI, drug delivery, magnetic separation and immobilization of biosubstances, *Biotechnol. Adv.*, **33**, pp. 1162–1176.

Holzwarth, U., Gibson, N. (2011). The Scherrer equation versus the 'Debye-Scherrer equation', *Nature Nanotechnol.*, **6**, p. 534.

Hu, L., Cao, X., Li, M., Hong, H., Xu, Q., Ge, J., Wang, L., Lu, J., Chen, L., Gu, H. (2011). Oxidation of benzylic compounds by gold nanowires at 1 atm O_2, *Chem. Commun.*, **47**, pp. 1303–1305.

Huang, C., Wen, T., Shi, F. J., Zeng, X. Y., Jiao, Y. J. (2020). Rapid detection of IgM antibodies against the SARS-CoV-2 virus via colloidal gold nanoparticle-based lateral-flow assay, *ACS Omega*, 5, pp. 12550–12556.

Hubert, V., Dumot, C., Ong, E., Amaz, C., Canet-Soulas, E., Chauveau, F., Wiart, M. (2019). MRI coupled with clinically-applicable iron oxide nanoparticles reveals choroid plexus involvement in a murine model of neuroinflammation, *Sci. Rep.*, **9**, p. 10046.

Ihs, A., Liedberg, B. (1991). Chemisorption of l-cysteine and 3-mercaptopropionic acid on gold and copper surfaces: an infrared reflection-absorption study, *J. Colloid Interface Sci.*, **144**, pp. 282–292.

Kievit, F. M., Wang, F. Y., Fang, C., Mok, H., Wang, K., Silber, J. R., Ellenbogen, R. G., Zhang, M. (2011). Doxorubicin loaded iron oxide nanoparticles overcome multidrug resistance in cancer *in vitro*, *J. Control. Release*, **152**, pp. 76–83.

Kim, B. H., Lee, N., Kim, H., An, K., Park, Y. I., Choi, Y., Shin, K., Lee, Y., Kwon, S. G., Na, H. B., Park, J. G., Ahn, T. Y., Kim, Y. W., Moon, W. K., Choi, S. H., Hyeon, T. (2011). Large-scale synthesis of uniform and extremely small-sized iron oxide nanoparticles for high-resolution T_1 magnetic resonance imaging contrast agents, *J. Am. Chem. Soc.*, **133**, pp. 12624–12631.

Kim, E.-Y., Stanton, J., Korber, B. T. M., Krebs, K., Bogdan, D., Kunstman, K., Wu, S., Phair, J. P., Mirkin, C. A., Wolinsky, S. M. (2008). Detection of HIV-1 p24 Gag in plasma by a nanoparticle-based bio-barcode-amplification method, *Nanomedicine*, **3**, pp. 293–303.

Lachowicz, D., Szpak, A., Malek-Zietek, K. E., Kepczynski, M., Muller, R. N., Laurent, S., Nowakowska, M., Zapotoczny, S. (2017). Biocompatible and fluorescent superparamagnetic iron oxide nanoparticles with superior magnetic properties coated with charged polysaccharide derivatives, *Colloid Surf. B Biointerfaces*, **150**, pp. 402–407.

Li, P., Liu, J. H., Yiu, H. P., Chan, K. K. (1997). Surface functionalization of polymer latex particles. I. Catalytic oxidation of poly(methylstyrene) latex particles in the presence of an anionic surfactant, *J. Polym. Sci. A Polym. Chem.*, **35**, pp. 1863–1872.

Lv, H., Zhang, S., Wang, B., Cui, S., Yan, J. (2006). Toxicity of cationic lipids and cationic polymers in gene delivery, *J. Control. Release*, **114**, pp. 100–109.

Malekzad, H., Hasanzadeh, M., Shadjou, N., Jouyban, A. (2017). Highly sensitive immunosensing of prostate specific antigen using poly cysteine caped by graphene quantum dots and gold nanoparticle: a novel signal amplification strategy, *Int. J. Biol. Macromolecule*, **105**, pp. 522–532.

Maneewattanapinyo, P., Banlunara, W., Thammacharoen, C., Ekgasit, S., Kaewamatawong, T. (2011). An evaluation of acute toxicity of colloidal silver nanoparticles, *J. Vet. Med. Sci.*, **73**, pp. 1417–1423.

Mazumder, V., Chi, M., More, K. L., Sun, S. (2010). Core/Shell Pd/FePt nanoparticles as an active and durable catalyst for the oxygen reduction reaction, *J. Am. Chem. Soc.*, **132**, pp. 7848–7849.

Mignot, C. (2018). Color (and 3D) for scanning electron microscopy, *Microsc. Today*, **26**, pp. 12–17.

Morana, G., Salviato, E., Guarise, A. (2007). Contrast agents for hepatic MRI. *Cancer Imaging*, **7**, (Special issue A), pp. S24–S27.

Morris, A. S., Adamcakova-Dodd, A., Lehman, S. E., Wongrakpanich, A., Thorne, P. S., Larsen, S. C., Salem, A. K. (2016). Amine modification of nonporous silica nanoparticles reduces inflammatory response following intratracheal instillation in murine lungs, *Toxicol. Lett.*, **241**, pp. 207–215.

Mourdikoudis, S., Menelaou, M., Fiuza-Maneiro, N., Zheng, G., Wei, S., Pérez-Juste, J., Polavarapu, L., Sofer, Z. (2022). Oleic acid/oleylamine ligand pair: a versatile combination in the synthesis of colloidal nanoparticles. *Nanoscale Horiz.*, **7**, pp. 941–1015.

Muthiah, M., Park, I.-K., Cho, C.-S. (2013). Surface modification of iron oxide nanoparticles by biocompatible polymers for tissue imaging and targeting, *Biotechnol. Adv.*, **31**, pp. 1224–1236.

Nirmal, M., Murray, C. B., Norris, D. J., Bawendi, M. C. (1993). Size-dependent spectroscopy and photodynamics of some ti-vt semiconductor nanocrystallites (quantum dots), *Proc. SPIE - Int. Soc. Opt. Eng.*, **1861**, pp. 280–284.

Parveen, S., Misra, R., Sahoo, S. K. (2012). Nanoparticles: a boon to drug delivery, therapeutics, diagnostics and imaging, *Nanomedicine: Nanotechnol. Biol. Med.*, **8**, pp. 147–166.

Petryk, A. A., Giustini, A. J., Gottesman, R. E., Trembly, B. S., Hoopes, P. J. (2013). Comparison of magnetic nanoparticle and microwave hyperthermia cancer treatment methodology and treatment effect in a rodent breast cancer model, *Int. J. Hyperth.*, **29**, pp. 819–827.

Pirkkalainen, K., Leppänen, K., Vainio, U., Webb, M. A., Elbra, T., Kohout, T., Nykänen, A., Ruokolainen, J., Kotelnikova, N., Serimaa, R. (2008). Nanocomposites of magnetic cobalt nanoparticles and cellulose, *Eur. Phys. J. D*, **49**, pp. 333–342.

Podgórski, R., Wojasiński, M., Ciach, T. (2022). Nanofibrous materials affect the reaction of cytotoxicity assays, *Sci. Rep.*, **12**, 9047.

Pokharkar, V., Dhar, S., Bhumkar, D., Mali, V., Bodhankar, S., Prasad, B. L. (2009). Acute and subacute toxicity studies of chitosan reduced gold nanoparticles: a novel carrier for therapeutic agents, *J. Biomed. Nanotechnol.*, **5**, pp. 233–239.

Raichlin, S., Sharon, E., Freeman, R., Tzfati, Y., Willner, I. (2011). Electron-transfer quenching of nucleic acid-functionalized CdSe/ZnS quantum dots by doxorubicin: A versatile system for the optical detection of DNA, aptamer-substrate complexes and telomerase activity, *Biosens. Bioelectron.*, **26**, pp. 4681–4689.

Roca, A. G., Costo, R., Rebolledo, A. F., Veintemillas-Verdaguer, S., Tartaj, P., González-Carreño, T., Morales, M. P., Serna, C. J. (2009). Progress in the preparation of magnetic nanoparticles for applications in biomedicine, *J. Phys. D: Appl. Phys.*, **42**, 224002.

Rodríguez-Luccioni, H. L., Latorre-Esteves, M., Méndez-Vega, J., Soto, O., Rodríguez, A. R., Rinaldi, C., Torres-Lugo, M. (2011). Enhanced reduction in cell viability by hyperthermia induced by magnetic nanoparticles, *Int. J. Nanomed.*, **6**, pp. 373–380.

Roth C. G. (2016). Fundamentals of Body MRI, 2nd ed. (Elsevier Saunders, USA).

Ruan, W., Zhou, T., Hui, G., Wang, Y., Chong, X., Wang, X., Song, W., Han, X., Zhao, B. (2014). Particle lithography-based patterning of polyelectrolyte template films and their application in fabrication of gold/silver nanoparticle assembly, *J. Colloid Interface Sci.*, **432**, pp. 65–69.

Sanz, C. G., Crisan, D. N., Leote, R. J. B., Onea, M., Barsan, M. M. (2022). Bioconjugates of mercaptocarboxylic acids functionalized AuNP and superoxide dismutase for superoxide electrochemical monitoring, *Microchim. Acta*, **189**, p. 245.

Sau, T. K., Pal, A., Pal, T. (2001). Size regime dependent catalysis by gold nanoparticles for the reduction of eosin, *J. Phys. Chem. B*, **105**, pp. 9266–9272.

Scarberry, K. E., Dickerson, E. B., McDonald, J. F., Zhang, Z. J. (2008). Magnetic nanoparticle-peptide conjugates for *in vitro* and *in vivo* targeting and extraction of cancer cells, *J. Am. Chem. Soc.*, **130**, pp. 10258–10262.

Schaffer, D. V., Koerber, J. T., Lim, K.-I. (2008). Molecular engineering of viral gene delivery vehicles, *Annu. Rev. Biomed. Eng.*, **10**, pp 169–194.

Scherer, F., Anton, M., Schillinger, U., Henke, J., Bergemann, C., Krüger, A., Gänsbacher, B., Plank, C. (2002). Magnetofection: enhancing and targeting gene delivery by magnetic force *in vitro* and *in vivo*, *Gene Ther.*, **9**, pp. 102–109.

Scherrer, P. (1918). Bestimmung der Größe und der inneren Struktur von Kolloidteilchen mittels Röntgenstrahlen, *Nachr. Ges. Wiss. Göttingen, Math.-Phys. Kl.*, **2**, pp. 98–100.

Sendroiu, I. E., Mertens, S. F. L., Schiffrin, D. J. (2006). Plasmon interactions between gold nanoparticles in aqueous solution with controlled spatial separation, *Phys. Chem. Chem. Phys.*, **8**, pp. 1430–1436.

Singh, P., Pandit, S., Mokkapati, V., Garg, A., Ravikumar, V., Mijakovic, I. (2018). Gold nanoparticles in diagnostics and therapeutics for human cancer, *Int. J. Mol. Sci.*, **19**, 1979.

Spirin, M. G., Brichkin, S. B., Yushkov, E. S., and Razumov V. F. (2020). Effect of surfactants on shape of gold nanoparticles, *High Energy Chem.* **54**, pp, 308–315.

Su, S., Fan, J., Xue, B., Yuwen, L., Liu, X., Pan, D., Fan, C., Wang, L. (2014). DNA-conjugated quantum dot nanoprobe for high-sensitivity fluorescent detection of DNA and micro-RNA, *ACS Appl. Mater. Interface*, **6**, pp. 1152–1157.

Sun, S., Anders, S., Hamann, H. F., Thiele, J-U., Baglin J. E. E., Thomson, T., Fullerton, E. E., Murray, C. B., Terris, B. D. (2002). Polymer mediated self-assembly of magnetic nanoparticles, *J. Am. Chem. Soc.,* **124**, pp. 2884–2885.

Sun, Y. H., Liu, Y. S., Vernier, P. T., Liang, C. H., Chong, S. Y., Marcu, L., Gundersen, M. A. (2006). Photostability and pH sensitivity of CdSe/ZnSe/ZnS quantum dots in living cells, *Nanotechnology*, **17**, pp. 4469–4476.

Takahashi, M., Akiyama, Y., Dcezumi, J., Nagata, T., Yoshino, T., Iizuka, A., Yamaguchi, K., Matsunaga, T. (2009). Magnetic separation of melanoma-specific cytotoxic T lymphocytes from a vaccinated melanoma patient's blood using MHC/peptide complex-conjugated bacterial magnetic particles, *Bioconjug. Chem.,* **20**, pp. 304–309.

Terán Arce, F., Vela, M. E., Salvarezza, R. C., Arvia, A. J. (1997). Comparative study of thiol films on C(0001) and Au(111) surfaces by scanning probe microscopy, *Surf. Rev. Lett.,* **4**, pp. 637–649.

Thoidingjam, S., Tiku, A. B. (2017). New developments in breast cancer therapy: role of iron oxide nanoparticles, *Adv. Nat. Sci.: Nanosci. Nanotechnol.,* **8**, p. 023002.

Tkachenko, A. G., Xie, H., Coleman, D., Glomm, W., Ryan, J., Anderson, M. F., Franzen, S., Feldheim, D. L. (2003). Multifunctional gold nanoparticle-peptide complexes for nuclear targeting, *J. Am. Chem. Soc.,* **125**, pp. 4700–4701.

Tomczak, N., Jańczewski, D., Han, M., Vancso, G. J. (2009). Designer polymer-quantum dot architectures, *Prog. Polym. Sci.,* **34**, pp. 393–430.

Torchilin, V. P. (2014). Multifunctional, stimuli-sensitive nanoparticulate systems for drug delivery, *Nat. Rev. Drug Discov.,* **13**, pp. 813–827.

Torres Sanchez, R. M., Ueda, A., Tanaka, K., Haruta, M. (1997). Selective oxidation of CO in hydrogen over gold supported on manganese oxides, *J. Catal.,* **168**, pp. 125–127.

U.S Food & Drug Administration (2016). Information on gadolinium-based contrast agents, https://www.fda.gov/drugs/postmarket-drug-safety-information-patients-and-providers/information-gadolinium-based-contrast-agents (accessed 31 August 2022).

Vergauwen, E. (2018). Central nervous system gadolinium accumulation in patients undergoing periodical contrast MRI screening for hereditary tumor syndromes, *Hered. Cancer Clin. Pract.,* **16**, pp. 1–9; Waiskopf, N., Shweky, I., Lieberman, I., Banin, U., Soreq, H. (2011). Quantum dot labeling of butyrylcholinesterase maintains substrate and inhibitor interactions and cell adherence features, *ACS Chem. Neurosci.,* **2**, pp. 141–150.

Waiskopf, N., Shweky, I., Lieberman, I., Banin, U., Soreq, H. (2011). Quantum dot labeling of butyrylcholinesterase maintains substrate and inhibitor interactions and cell adherence features, *ACS Chem. Neurosci.*, **2**, pp. 141–150.

Wang, H.-Q., Wang, J.-H., Li, Y.-Q., Li, X.-Q., Liu, T.-C., Huang, Z.-L., Zhao, Y.-D. (2007a). Multi-color encoding of polystyrene microbeads with CdSe/ZnS quantum dots and its application in immunoassay, *J. Colloid Interface Sci.*, **316**, pp. 622–627.

Wang, X., Zhang, R., Wu, C., Dai, Y., Song, M., Gutmann, S., Gao, F., Lv, G., Li, J., Li, X., Guan, Z., Fu, D., Chen, B. (2007b). The application of Fe_3O_4 nanoparticles in cancer research: a new strategy to inhibit drug resistance, *J. Biomed. Mater. Res. - A*, **80**, pp. 852–860.

Wierucka, M., Biziuk, M. (2014). Application of magnetic nanoparticles for magnetic solid-phase extraction in preparing biological, environmental and food samples, *TrAC - Trend Anal. Chem.*, **59**, pp. 50–58.

Wuister, S. F., Van Driel, F., Meijerink, A. (2003). Luminescence and growth of CdTe quantum dots and clusters, *Phys. Chem. Chem Phys.*, **5**, pp. 1253–1258.

Xiao, Y. D., Paudel, R., Liu, J., Ma, C., Zhang, Z. S., Zhou, S. K. (2016). MRI contrast agents: classification and application (Review), *Int. J. Mol. Med.*, **38**, pp. 1319–1326.

Yan, F. (2013). Plasmonic colorimetry: visual detection of environmentally relevant species, *Int. J. Environ. Technol. Manag.*, **16**, pp. 49–64.

Yang, P.-F., Lee, C.-K. (2007). Hyaluronic acid interaction with chitosan-conjugated magnetite particles and its purification, *Biochem. Eng. J.*, **33**, pp. 284–289.

Yiu, H. H. P. (2011). Engineering the multifunctional surface on magnetic nanoparticles for targeted biomedical applications: a chemical approach, *Nanomedicine*, **6**, pp. 1429–1446.

Yu, T., Greish, K., McGill, L. D., Ray, A., Ghandehari, H. (2012). Influence of geometry, porosity, and surface characteristics of silica nanoparticles on acute toxicity: their vasculature effect and tolerance threshold, *ACS Nano*, **6**, pp. 2289–2301.

Zadnik, P. L., Molina, C. A., Sarabia-Estrada, R., Groves, M. L., Wabler, M., Mihalic, J., McCarthy, E. F., Gokaslan, Z. L., Ivkov, R., Sciubba, D. (2014). Characterization of intratumor magnetic nanoparticle distribution and heating in a rat model of metastatic spine disease, *J. Neurosurg. Spine*, **20**, pp. 740–750.

Chapter 2

Ordered Mesoporous Silicas: Liquid Crystal Templating

Humphrey H. P. Yiu
School of Engineering and Physical Sciences, Heriot-Watt University, Edinburgh, EH14 4AS, UK
h.h.yiu@hw.ac.uk

2.1 The Discovery of Ordered Mesoporous Silicas

In Chapter 1, the use of surfactants and micelles as templates for making nanoparticles has already been introduced. We may consider such nanoparticles as 0-D (zero dimension) nanomaterials, while nanorods or nanowires, including carbon nanotubes, are usually referred to as 1-D nanomaterials. How about 2-D or 3-D nanomaterials? Nanomaterials of a higher complexity can be made by exploiting another property of surfactants and micelles; the formation of *liquid crystals*. Liquid crystal templating provides a method of preparing highly ordered, 3-D nanoporous materials. In this chapter, we will focus on the synthesis, properties, and applications of these delicate nanomaterials.

Ordered nanoporous materials may sound like some very recent inventions. In fact, scientists have been studying these materials for over 200 years. Pillared clay minerals are classic examples of

Nanotemplating: Engineering Nanocomposite Materials for the Future
Humphrey H. P. Yiu

ISBN 978-981-4316-92-7 (Hardcover), 978-1-003-50999-8 (eBook)
www.jennystanford.com

an ordered porous structure built with sheets of aluminosilicate, with a thickness of only a few nanometers. Another family of natural aluminosilicates, called zeolites, have 3-D interconnected nanoporous structures. Interestingly, both clays and zeolites can be found naturally. Although zeolites and clay materials have nanoporous structures, using them as hard templates for making other nanostructured materials with larger porous structures is unlikely due to different reasons. The 2-D sheet structures of pillared clays (templates) simply cannot be retained during the casting process. In the case of zeolites, the pores are simply too small (mostly ≤ 1 nm) to create any nanostructured materials strong enough for further applications after the removal of the template. Also, the small pores of zeolites limit the variety of precursor molecules that can be used for making the nanostructured materials. Scientists need some more versatile methods for building ordered nanoporous materials with larger pore dimensions.

Before the 1990s, scientists and industries had been using zeolites and acid-treated clays as solid acid catalysts for many industrial processes, including hydrocarbon cracking of large hydrocarbons (alkanes) in petroleum industries, alkylation and trans-esterifications (e.g., Primo and Garcia, 2014). However, the pores in zeolites were still too small to accommodate many organic molecules, while using pillared clays was simply too problematic with impurities and poor quality control. Synthesizing large-pore zeolites had become a holy grail in inorganic materials chemistry in the 1980s.

The breakthrough happened in 1992 when a research group at Mobil USA reported their version of ordered mesoporous silica, MCM-41 and MCM-48 (Mobil Composite Materials) using a cationic surfactant as a template (Kresge et al., 1992). It took a while for scientists to understand the formation mechanism of these materials, known as liquid crystal templating (LCT). Although some scientists still suspect that the discovery of MCM materials was a coincidence, these materials did open up an entirely new research direction for solid-state chemistry in the next two decades. To date, the original Mobil paper has already been cited over 15,000 times since 1992.

Because of their success, some scientists also wondered if the Mobil group was actually the first to report ordered mesoporous silica materials. It is believed that the first report on synthetic mesoporous material was patented in 1969 (Chiola et al., 1969). Because of poor analytical techniques in that era, full characterization of this material was never established. In 1990, hexagonally packed mesoporous silicas, prepared from layered sodium silicate, Kanamite with a surfactant template, were reported by Yanagisawa et al. (Yanagisawa et al., 1990). This material is called FSM (Folded Sheet Materials) and it was little known in the early 1990s (Fukushima and Inagaki, 1996). This report has only 13 citations in the first 5 years (1990–1994); this is dwarfed by the massive >300 citations for the Mobile paper in 5 years (1992–1996). These FSM materials were reported earlier than the MCM series but many research groups still cite the Mobil papers as the first report for ordered mesoporous silica.

Since then, following this liquid crystal templating route, scientists started to find more new mesoporous structures. The other widely used and frequently researched ordered mesoporous materials belonging to the *SBA* family originated from the University of California, Santa Barbara (hence SBA). The first SBA material was reported in 1996 (Huo et al., 1996) but the more widely used SBA-15 silica was reported in 1998 (Zhao et al., 1998a). Hundreds of other mesoporous materials were reported since but most of them were based on either the MCM or the SBA formulation.

In the 1990s, research on the applications of mesoporous silica was mainly directed to heterogeneous catalysis for commercial chemical products. This is the consequence of the original idea of making a "large-pore zeolite" during the discovery of MCM-41, driven by the chemical industry. Unfortunately, the amorphous nature of mesoporous silica leads to a lower catalytic activity, that can never match those observed from zeolites. As a result, scientists started to look at mesoporous silicas in a different way, not just as a large-pore zeolite, but as delicate nanostructures for specific purposes. Three main examples can illustrate these trends: (1) functionalization of mesoporous silica with organic groups; (2) using mesoporous silica in biomolecular research; and (3) using mesoporous silica as a template for other inorganic materials.

2.2 Characteristics of Ordered Mesoporous Silicas

2.2.1 Definition and Classification

Many silica materials (e.g., silica gels, fumed silica) exhibit mesoporous characters, with random disordered mesopores. However, here we refer ordered mesoporous silicas to as those silica materials that show a long-range order of pores in a meso-scaled range (typically 2–10 nm) and a narrow pore size distribution (within ±1 nm). Although they have long-range order porous systems and may even exist as large "crystals," these ordered mesoporous silicas are built with amorphous (disordered) "walls." This is a stark contrast to crystalline silica materials such as quartz. Ordered mesoporous silicas are essentially amorphous materials and their silica frameworks are rather flexible. The flexibility of silica frameworks is also one of the reasons that silica can form a number of mesophases with long-range orders while other (metal) oxides are less successful in doing so.

One class of disordered materials is commonly classified as mesoporous silicas. Mesocellular Foams (MCFs), formed by the addition of a swelling agent (e.g., trimethylbenzene or mesitylene) to an SBA-15 synthesis, show disordered mesoporous structure with pore sizes ranging from 7 to 10 nm. They are also commonly referred to as mesoporous silica due to their similarity in synthesis procedure and formation mechanism. Indeed, MCFs have been widely used as a material to compare with SBA-15 (MCF materials will be discussed in detail in Chapter 3). Another disordered material HMS (as hexagonal mesoporous molecular sieves) is also often referred to as mesoporous silica. Figure 2.1 shows a few commonly found structures in mesoporous silicas.

Classification of ordered mesoporous silica is rather confusing. This is because the notation of mesoporous silica has no reference to the ordered structure or the surfactant template used. Therefore, SBA-15 and SBA-2 have very few common features except the fact that they are both mesoporous silicas developed by a research group in Santa Barbara, USA; on the other hand, both SBA-16 and FDU-1 have a *Im3m* cubic symmetry and similar pore size but formed with different surfactants. Because they were developed by different research institutions, they have completely different, unrelated notations. Indeed, SBA-16 and FDU-1 were developed by the same scientist (Prof. Zhao) while working in different institutes (Yu et al., 2001). In general, the first three letters of the early developed mesoporous silica notation usually indicated the research institute first reporting the material, while the number suffix is likely to be the order of the development. For example, FDU-1 and FDU-5 were both developed at Fudan University in Shanghai, China, but FDU-5 was reported 2 years after FDU-1. We also see MCM-22 reported in the literature (Corma et al., 1995) but MCM-22 is a zeolite, not a mesoporous silica. This was developed before MCM-41 as the number suggests (Rubin and Chu, 1990). Indeed, one of the most misleading notations was HMS (referring to hexagonal mesoporous silica) prepared with hexadecylamine ($C_{16}H_{33}$-NH_2) as it has no proven hexagonal structure from the original paper (Tanev and Pinnavaia, 1996).

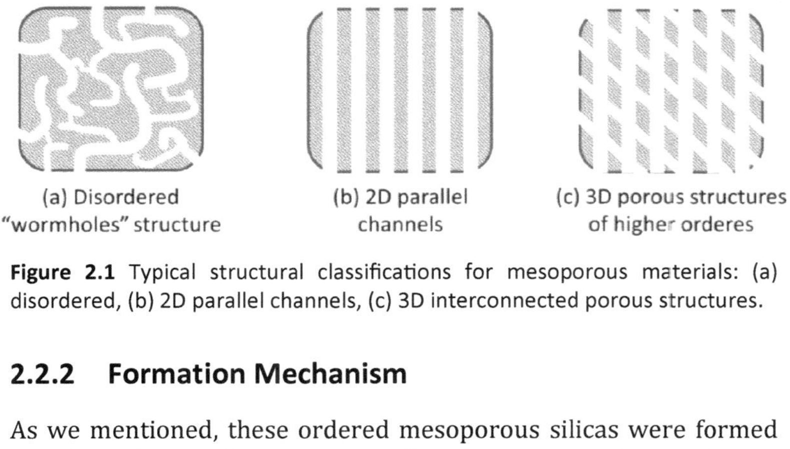

Figure 2.1 Typical structural classifications for mesoporous materials: (a) disordered, (b) 2D parallel channels, (c) 3D interconnected porous structures.

2.2.2 Formation Mechanism

As we mentioned, these ordered mesoporous silicas were formed via the LCT mechanism. Logically, their synthesis requires a "liquid crystal" and a silica source. However, at the beginning of mesoporous silica research, it was believed that these ordered mesoporous silicas were formed by coating the liquid crystal template with the silica source, followed by condensation/crystallization. Since the concentration of surfactant used in synthesis is always much lower than the "critical micelle concentration 2" or CMC-2, which is the minimum concentration of a surfactant to form a liquid crystal phase, the "coating" mechanism was dismissed. Nonetheless, it is possible to form ordered mesoporous silica by coating a preformed liquid crystal with a silica source, as demonstrated by Prof. George Attard, University of Southampton, U.K. (Attard et al., 1995). A much higher surfactant content is usually required if this route is chosen due to the high surfactant concentration required to achieve a CMC-2 phase.

Indeed, the formation of ordered mesoporous silica phase during synthesis was triggered by the addition of the silica source in a solution with surfactant micelles (step 2 in Fig. 2.2). Varying the synthesis conditions, e.g., pH, salt, co-solvent or co-surfactant, may give a different phase. Figure 2.2 illustrates a simplified formation mechanism for a 2-D hexagonal mesoporous silica.

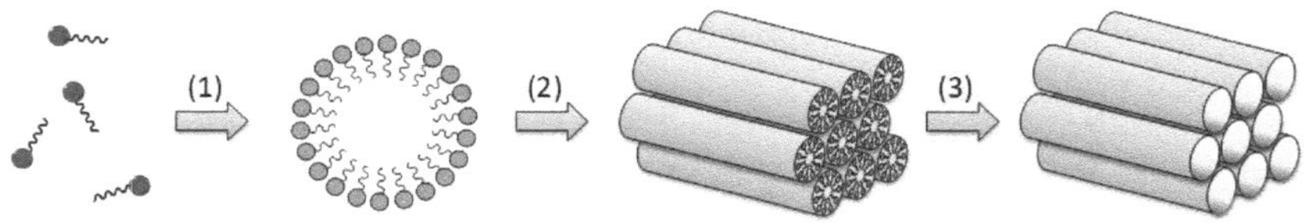

Figure 2.2 Formation mechanism of mesoporous silica via liquid crystal templating (LCT) route: step (1) formation of micelle at CMC-1, step (2) formation of liquid crystal triggered by the addition of a silica precursor, and step (3) removal of surfactant.

Table 2.1 Selected formation mechanisms of mesoporous silicas based on the surfactant/inorganic (S/I) interaction

S/I interaction	Mesoporous silica example	Surfactant used (I)	Silica source (S)
S^+I^-	MCM-41	CTAB	$NaSiO_3$
S^-I^+	[a]	SDS	APTES
S_0I_0	HMS	Hexadecylamine	TEOS
$S_0H^+X^-I_0$	SBA-15	$EO_{20}PO_{70}EO_{20}$ (P123)	TEOs

Note: The "+" and "–" signs indicate the charge of the species, while "0" denotes the neutral species.

[a]No specific notation was given to this mesoporous silica.

During the synthesis, the nature of interactions between surfactant (S) and inorganic substrates (I) is important. Generally, such S/I interactions can be classified according to their charges.

On many occasions, a third/fourth component(s) such as the pH regulator (e.g., H^+) and the counter ions (M^+ or X^-) could also take place in this interaction. Table 2.1 summarizes the four common S/I interactions for the synthesis of mesoporous silica.

2.2.2.1 Liquid crystal templating and phase diagram

Due to their LCT formation mechanism, naturally, understanding liquid crystals is essential to the preparation of ordered mesoporous materials. Usually, lyotropic liquid crystals (liquid crystals formed with solvent) are used for preparing ordered mesoporous silicas, instead of thermotropic liquid crystals (formed due to change in temperature) as the structure-directing reagent. In order to understand the different phases of mesoporous silicas, it is essential to start with the liquid crystal phase.

To determine the phase of a liquid crystal, a phase diagram can be used as a reference. However, the addition of an extra component into the reaction mixture can significantly alter the phase diagram and it can be difficult to predict the effect of this extra component. As a result, a new phase diagram must be established for a new system. Figure 2.3 shows a phase diagram of cetyltrimethylammonium bromide (CTAB) for the formation of MCM-41 and MCM-48. Such transition can also be seen with other surfactant systems.

According to Fig. 2.3, more than one liquid crystal phase can be formed from one surfactant simply by varying its concentration. In theory, this rule applies to every lyotropic surfactant but identifying these phases can be a very difficult task. Hence, only a handful of surfactants used for preparing ordered mesoporous silica have their phase diagram reported. Many scientists simply take advantage of the established phase diagram and make small changes in the preparation to make new structures. One commonly used method is the addition of oil (such as mesitylene) to increase the pore dimension by swelling the liquid crystal. Though this expansion has its limit and over-expansion will cause the loss of order in the liquid crystal phase, leading to disordered materials. Another useful modification is the addition of a co-solvent (such as isopropanol) in order to modify the morphology of the final material.

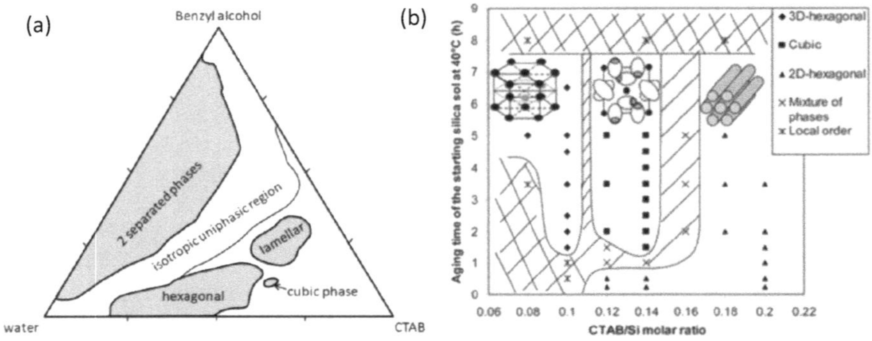

Figure 2.3 Liquid crystal phase diagrams of (a) CTAB (adapted from Montalvo et al., 1995) and (b) CTAB with silica source (Besson et al., 2003).

2.2.2.2 Mesostructures or mesophases

Ordered mesoporous silicas are uniquely characterized by their long-range ordered mesostructure (or mesophase). This is strongly linked to the liquid crystal phases. The Mobil's publication in 1992 on MCM materials has already reported that using the same surfactant, three mesophases can be achieved by varying reaction conditions; 2-D hexagonal MCM-41, 3-D cubic MCM-48 and lamellar MCM-50. Since then, many different mesophases have been reported. In general, there are four main classes of mesostructures: hexagonal, cubic, lamellar and disordered (or "wormholes"). Table 2.2 highlighted a number of common mesoporous silica with their mesophases as well as symmetry.

2.2.3 Inorganic Substitution on Framework

We should note that all materials presented in Table 2.2 are all purely siliceous. Pure silica by itself is relatively inert in terms of chemistry. Therefore, apart from the separation of molecules based on size exclusion, these mesoporous silicas do not have many applications without modification. Many modification pathways have been reported for mesoporous silicas in order to incorporate some chemical functionalities onto the materials. Vaguely, we can group these methods into (1) inorganic substitution, and (2) silanization to introduce functional organic groups.

Table 2.2 Selected examples of common mesoporous silicas that were prepared from liquid crystal templating with their structural symmetry, surfactant (S) and silica sources (I) that were used

Mesophase	Mesoporous silica example	Symmetry	Surfactant used (S)	Silica source (I)
Hexagonal	FSM-16	p6mm	CTAB	$NaSiO_3$
	MCM-41	p6mm	CTAB	$NaSiO_3$
	SBA-15	p6mm	$EO_{20}PO_{70}EO_{20}$ (P123)	TEOS
	SBA-2	$p6_3/mmc$	divalent surfactants	TEOS
	SBA-12	$p6_3/mmc$	$C_{18}EO_{10}$	TEOS
Cubic	MCM-48	Ia3d	CTAB	TEOS
	KIT-6	Ia3d	$EO_{20}PO_{70}EO_{20}$ (P123)	TEOS
	FDU-5	Ia3d	$EO_{20}PO_{70}EO_{20}$ (P123)	TEOS
	SBA-1	Pm3n	$C_nH_{2n+1}(C_2H_5)_3N^+$	TEOS
	SBA-16	Im3m	$EO_{100}PO_{69}EO_{100}$ (F127)	TEOS
	FDU-1	Im3m	$EO_{100}PO_{69}EO_{100}$ (F127)	TEOS
Lamellar	MCM-50	L	CTAB	TEOS
Disordered	HMS	None	Hexadecylamine	TEOS

For inorganic substitution, usually silicon atoms in the silica framework are replaced by metal atoms. There are two main objectives to carry out substitutions with metal on mesoporous silicas; to generate a charged surface or to incorporate a chemically active element. Due to the differences in atomic valencies (i.e., how many electrons are there on the atom's outmost shell), substitution can generate a charged surface. For example, valency on silicon atoms in a silica framework can be considered as +4, i.e., 4 electrons are required from oxygen to make them neutral. If a silicon atom was substituted by another atom of +3 in valency (e.g., aluminum, Al), the surface will become negatively charged. This charged surface results in some alteration of the chemical properties of the materials. In order to balance the charge, hydrogen (H^+) ions can be adapted for the purpose and these H^+ ions are acidic, giving the acidic properties to aluminosilicates, as seen in

Al-substituted mesoporous silicas. Moreover, these H^+ ions are also exchangeable with other cation ions, e.g., Na^+ of Ca^{2+}, acting as a cation exchanger. Such behavior also resembles that observed on zeolites.

In other examples, this substitution can involve elements that have some intrinsic properties, producing a new material carrying new properties due to substitution. Titanium (Ti) substitution is a good example of introducing the oxidative property to mesoporous silica materials. The use of these metal-substituted mesoporous silicas will be discussed in detail in Section 2.4.

2.2.4 Periodic Mesoporous Organosilicas (PMOs)

In addition to metal substitution, organic functional groups can also be attached to the siliceous wall of mesoporous silica. Inorganic substitution on mesoporous silica can be compared with those in zeolites. However, it is rather rare to have conventional zeolites (except metal-organic frameworks or MOFs, which are now commonly classified as a separate family of materials) with organic functional groups inside pores due to the limited space available for hosting organic functionalities. For example, a simple benzene ring is around 0.36 nm wide while the pore of zeolite A is just marginally larger at 0.42 nm. (see Fig. 2.4) This is hardly a good fit. Moreover, even if it fits, these functional groups have already occupied most of the space in zeolite pores, leaving no room for substrates to enter. As a result, organically functionalized zeolite is rare. The larger pores of mesoporous silica become much more adequate. Another issue is that zeolites are crystalline in microstructure while mesoporous silicas are amorphous. These organic groups can attach to a more flexible framework via a Si-C linkage. Therefore, the organically functionalized mesoporous silica can be very stable at room conditions.

Organic functionalities can also be incorporated into the ordered mesoporous silica structures; they have been termed "periodic mesoporous organosilicas" or PMOs. However, many research papers do not use this term since this class of materials is very diverse. Most of the organic functional groups are introduced using organosilanes. Organosilanes are silicon compounds with at least one Si–C linkage, to provide the organic functionality. One classic example of organosilane for functionalizing mesoporous silica is 3-aminopropyltriethoxysilane, or APTES. Figure 2.5 shows the structure of APTES; many other organosilanes have a similar structure but with a critical difference at the end of the short carbon chain, the functional group "X." There are a wide variety of "X" groups, e.g., alkyl chain C_nH_{2n+1}, Br, Cl, CN, SH, epoxy, and PEG, etc. Other organosilanes, including trimethoxy- and trichloro-, instead of triethoxy-, are also available. These varieties hydrolyze or decompose in a similar manner.

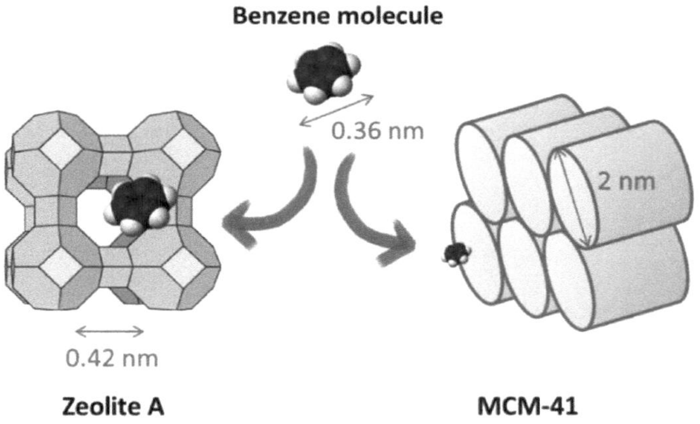

Figure 2.4 An illustration comparing the pore dimension of zeolite A and MCM-41 to a benzene molecule.

Amine functionalization using APTES is one of the earliest to be developed. (Macquarrie, 1996) There are several reasons; APTES offers an amine surface on mesoporous silica and amine is rather easily activated. It allows for a second step further functionalization. Moreover, the amine is a positively charged group, giving the silica surface a contrast in the surface property, as unfunctionalized silica is slightly negatively charged. Finally, the decomposition reaction of these organo(m)ethoxysilanes is catalyzed by a base. APTES (or APTMS for methoxysilane) can also act as a "self-catalyst" due to the basicity of the amine group.

Therefore, the efficiency in functionalization using APTES is generally high. All these reasons combined, the functionalization of mesoporous silica using APTES has been very popular ever since.

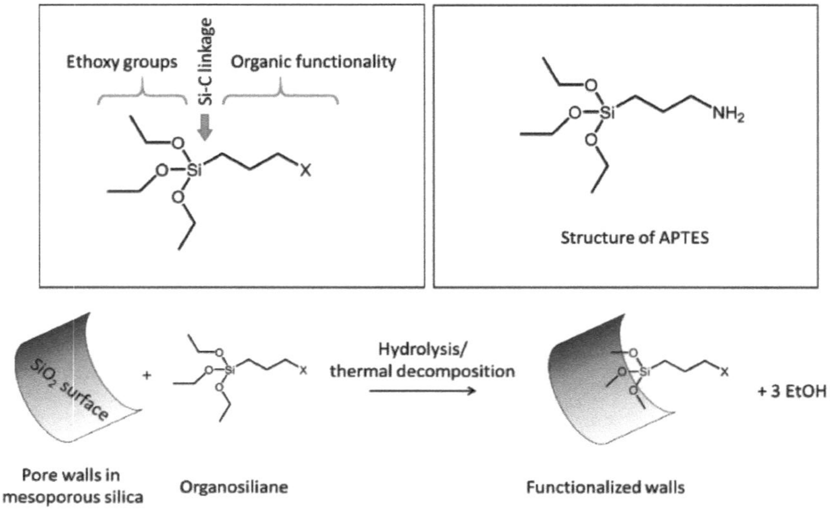

Figure 2.5 Structure of a typical organosilane for surface functionalization of mesoporous silica, structure of APTES and a typical scheme for surface functionalization of mesoporous silica.

There are two commonly used methods for incorporating organic functionalities onto the mesoporous silica surface; (1) *in situ* co-condensation and (2) post-synthesis grafting.

2.2.4.1 *In situ* one-pot co-condensation

In situ co-condensation functionalization was first reported in 1996 (Burkett et al., 1996; Macquarrie, 1996). The principle is very simple as *in situ* co-condensation requires both the silica source (usually TEOS) and the organosilane (e.g., APTES) in the same synthesis step before hydrolysis starts. Indeed, these two species will hydrolyze simultaneously once the reaction starts. Once the mesoporous silica solid is formed, the designated organic functional groups will be integrated randomly into the structure. The surfactant templates are required to be removed via solvent extraction. Calcination will only burn the organic groups. Once the pores are free of surfactant, this organic functionalized mesoporous silica is ready for use.

2.2.4.2 Post-synthesis grafting

Organosilanes can also be grafted after the mesoporous silicas are fully formed/calcined. Grafting organic groups onto mesoporous silicas was reported before the *in situ* co-condensation route, (Brunel et al., 1995) but the earlier reported methods were rather complicated, needing to activate the silica surface first. One early example of one-step grafting using organosilanes was reported in 1996 where a chlorosilane was used to introduce alkyl chloride groups on an MCM-41 surface (Sutra and Brunel, 1996). Further functionalization (grafting a manganese(III) Schiff-base complex) was then carried out. Since then, the use of organosilanes became a preferred route for post-synthesis grafting, partly due to the wide range of organosilanes being commercially available for many different purposes.

In a typical grafting experiment, an unfunctionalized mesoporous silica sample is suspended in a solvent (usually dry toluene) and the organosilane is added to the reaction system at around 1% concentration (e.g., 1 g silane in 100 mL toluene). The reaction can then heat up to a reflux (110°C for toluene) for 8–16 h. Since the amount of silanes used is usually in excess, the functionalized mesoporous silica needs to be washed with a volatile solvent (e.g., acetone) to remove the unreacted silanes and the excessive solvent. After drying, the post-synthesis functionalized mesoporous silica samples are ready for use.

2.2.4.3 Comparison of *in situ* vs post-synthesis method

Both of these two methods for the organic functionalization of mesoporous silicas are used often in research. They both have their advantages and disadvantages. Therefore, the choice is highly dependent on the end use of the materials and the available facilities in the laboratory. *In situ* one-pot preparation allows a one-step synthesis and the co-condensation process is regarded as a route to produce more robust materials with organic groups firmly anchored onto the silica framework. However, the amount of the organosilane used is limited to ~ 5% (e.g., TEOS:organosilane = 19:1); higher organic content will lead to loss of the long-range order, which is characteristic for ordered mesoporous silica. In some cases, the introduction of an organosilane can alter the mesophase and the effect is difficult to predict. Also, the materials have to undergo template extraction using ethanol or other organic solvent. This is a time-consuming process and the materials will also be left with some surfactant residue. Complete extraction is not common. Finally, the homogeneity of the materials depends on the rate of hydrolysis but silanes with different functional groups have different hydrolysis rates. For example, APTES hydrolyzes faster due to its amine group, which is capable of self-catalyzing the hydrolysis step. Besides, we may also have to mix ethoxysilane with an organo-functionalized methoxysilane, depending on the availability. This will further increase the heterogeneity of the final material.

On the other hand, post-synthesis grafting allows full calcination so no surfactant residues will be left in the materials. The silanization step, at 110°C, is unlikely to alter the mesophase of the host material. Nonetheless, this is a two-step process (synthesis of mesoporous silica, followed by silanization), and the stability of the grafted organic groups depends on the grafting step; the organosilanes may not be fully hydrolyzed, reducing the group stability. The earlier reports seemed to have recorded some leaching or functional groups from the post-synthesis method (Macquarrie et al., 1999). The available silanol groups on the surface for grafting may be reduced after calcination; regeneration of these silanol groups using an HCl pretreatment has been suggested but not commonly used. High organic content could also lead to pore blockage. Finally, for some organic chemistry laboratories, a furnace for calcination may not be available; hence post-synthesis could be less attractive.

2.3 Preparation of Ordered Mesoporous Silica

The preparation procedure for ordered mesoporous silicas varies depending on which literature was referred to. However, there are a handful of essential ingredients for the formation of these nanomaterials: a surfactant template, a silica source, a pH regulator and water. Other ingredients can be added in order to modify the structure or the composition of the final product.

2.3.1 Synthesis Procedure of Ordered Mesoporous Silica

Preparation procedures of ordered mesoporous silicas are not complicated. Again, there are some variations in reaction temperature and time but, generally, it is a three-step synthesis. These three steps are: preparing the gel, hydrothermal treatment, and removal of the surfactant template.

To prepare the gel, the surfactant needs to be dissolved in water together with the pH regulator, usually HCl or NaOH. Once all surfactant is dissolved, the silica source can be added slowly with constant stirring. It is essential to ensure all surfactant is dissolved before adding the silica source. Otherwise, a mixed phase material may form as the surfactant concentration is constantly changing until complete dissolution. Usually, this gelling process is carried out at room temperature but, for some protocols, this needs to be done at a higher temperature (40°C for SBA-16 and 30°C for SBA-15) or a lower temperature (0°C for SBA-2). Any additional modifier for pore structure or morphology should be added prior to the addition of a silica source.

In many cases, the gel needs to be submitted to hydrothermal treatment in order to increase the "crystallinity" or the long-range order of the mesoporous silica. The hydrothermal process is commonly used for the preparation of many crystalline metal oxides. During a hydrothermal process, the gel is heated up at 100–200°C to create a pressurized environment and encourage crystallization. For ordered mesoporous silica, the gel is transferred to a sealed PTFE bottle or even a PTFE-lined stainless steel autoclave. The choice of the vessel depends on the temperature required. PTFE bottle is adequate for a temperature up to 150°C and an autoclave is necessary for a higher temperature. Maximum operation temperature is usually available from the manufacturer's guideline. This hydrothermal process lasts from 24 h to 4 days. It is worth noting that the high pressure created inside the vessel can be dangerous when the mesoporous silica is to be recovered after the hydrothermal treatment. The vessel is usually cooled down before releasing the pressure. Extra caution is needed when opening the vessel.

The use of pressurized vessels for the hydrothermal synthesis of mesoporous silica makes the production rather tricky to be scaled up. That is why mesoporous silica materials are relatively expensive commercially. For example, in 2022, the price of SBA-15 is 167 USD for 5 g, while fumed silica costs 92 USD for 500 g from Sigma Aldrich. Other MCM and SBA silicas are even more expensive, at up to 400 USD for 5 g. (Sigma Aldrich, 2022). There are several reports on the preparation of mesoporous silica at ambient pressure, but the product tends to be less crystalline (Setoguchi et al., 1997). Nonetheless, the quality of the ordered mesoporous silicas made at ambient pressure is improving.

One recent direction of research is to develop the continuous synthesis of ordered mesoporous silicas. In the chemical industry, continuous processes are considered time- and labor-saving compared with batch syntheses. This could drive down the manufacturing cost, making them more commercially acceptable. There has been some success in this development, but continuous synthesis is still not widely used for preparing ordered mesoporous silicas (Ryu et al., 2019).

Like many solid–liquid reactions, the mesoporous silica is recovered by simple filtering. The solid should be washed with water to remove unwanted surfactants and other residues. Ethanol may be used to make the solid dry quicker. Once the mesoporous silica is dried, the surfactant template inside the pore needs to be removed so that the network of nanopores can be revealed. There are two methods to remove these surfactants: (1) by calcination (burning in air) or (2) by extraction.

Calcination is a straightforward method to remove the surfactant template inside the mesoporous silica. Usually, the recovered solid is submitted to a furnace and heated to 550°C in air for several hours. Ideally, the calcination process is to be done inside the fume cupboard as fume with a strong smell, due to some volatile organic compounds, is discharged during heating.

This problem is particularly severe for the calcination of MCM-48, which is made with a high concentration of ammonium surfactant. The use of pure oxygen during calcination may cause a minor explosion due to the heat release and the high surfactant content in the as-prepared mesoporous silica. In this case, it is advised to heat the as-prepared mesoporous silica in flowing nitrogen to remove some volatile organics before switching to oxygen when the sample is cooled down. One should NEVER switch to oxygen when the furnace is still hot.

If calcination needs to be avoided in order to preserve the organic functionalities of as-prepared mesoporous silica, extraction is the only alternative method to remove the surfactant template. Ethanol is a good solvent for extraction because it dissolves many surfactants. In a typical extraction experiment, as-prepared mesoporous silica is suspended in ethanol and heated to a reflux for at least 8 h. To maximize the extraction, this process should be repeated several times. Despite that, 100% extraction can be difficult to achieve and some surfactant template may be left inside the pores. A few research groups had reported the use of supercritical CO_2 for the extraction of surfactant template for maximum extraction efficiency (Ghaedi and Zhao, 2022). However, this method has not been popular with scientists due to the high energy requirement for this process. Once the surfactant is removed, the ordered mesoporous silica is ready for use.

2.3.2 Characterizations of Ordered Mesoporous Silica

The synthesized mesoporous silicas should be fully characterized as the final materials may not always form as planned. Transmission electron microscopy (TEM) is an ideal technique to study the nanostructure of ordered mesoporous silicas with images of high magnifications (up to ×200k magnification would normally suffice). Scanning electron microscopy (SEM) provides information on particle morphology while powder x-ray diffraction (PXRD) depicts the long-range order of the template. Nitrogen adsorption isotherms also offer data on pore dimensions and the pore volume, which determines the amount of cast nanomaterials that can be formed.

Box 2.1

Synthesis for SBA-15

Chemicals:

Silica source: tetraethylorthosilicate TEOS (8.5 g)

Surfactant template: co-block polymer Pluronic P123 $EO_{20}PO_{70}EO_{20}$ (4.0 g)

pH regulator: conc. HCl (35% w/w, 12.5 g)

solvent: water (138 mL)

Procedure:

1. Pluronic P123 is dissolved in water by magnetic stirring in a polypropylene beaker (250 mL) at 40°C. HCl is added and stirred for at least an hour. Caution: Handling conc. HCl requires special care and protective gloves have to be worn.
2. When all surfactant is fully dissolved, TEOS is added dropwise while stirring. The reaction mixture is stirred at 40°C for 24 h.
3. A white colloidal suspension is formed after 24 h. This suspension is transferred to a Teflon bottle (250 mL) and heated to 100°C in an oven for 48 h.
4. The white solids are filtered, washed and dried in air.
5. Finally, the dry white powder is "fired" (or calcined) in a furnace at 550°C for at least 4 h to burn off all surfactant. Caution: pure oxygen is to be avoided for firing as an explosion may occur.
6. Characterizations (XRD, TEM, N_2 adsorption) are to be carried out for structural analysis.

Box 2.2

Synthesis for MCM-48

Chemicals:

Silica source: tetraethylorthosilicate TEOS (10.5 g)

Surfactant template: cetyltrimethylammonium bromide CTAB (11.88 g)

pH regulator: NaOH pellets (1.05 g)

solvent: water (150 mL)

Procedure:

1. CTAB is dissolved in water by magnetic stirring in a polypropylene beaker (250 mL) at room temperature. NaOH

is added and stirred for at least an hour. Caution: Handling NaOH requires special care and protective gloves have to be worn.

2. When all surfactant is fully dissolved, TEOS is added dropwise while stirring. The reaction mixture is stirred at room temperature for 24 h. N.B. At a high concentration of surfactant, the solution of CTAB may appear to be slightly cloudy.
3. A white colloidal suspension is formed after 24 h. This suspension is transferred to a Teflon bottle (250 mL) and heated to 100°C in an oven for 72 h.
4. The white solids are filtered, washed and dried in air.
5. Finally, the dry white powder is "fired" in a furnace at 550°C for at least 4 h to burn off all surfactant. Caution: pure oxygen is to be avoided for firing as an explosion may occur.
6. Characterizations (XRD, TEM, N_2 adsorption isotherm measurement) are to be carried out for structural analysis.

2.3.2.1 Transmission electron microscopy

The ordered structure of mesoporous silicas can be studied by TEM. With this method, the arrangement of the pores can be visually shown in a photograph. TEM is a powerful technique to examine the microstructure of a material. When ordered mesoporous silicas were first developed, dimensional images of samples can be observed on the viewing screen or printed as black and white photographs. Nowadays, we can study digital images with the help of a digital camera fitted inside the microscope. This has the advantage of taking a large number of images without laborious photograph development in a darkroom. Modern TEM instruments allow high resolutions of atomic level (termed as HRTEM) and the lattice of a solid can now be shown at an atomic level. There are two basic modes of images—bright field and dark field—to view a sample in an electron transmission microscope. Generally, the bright field technique is suitable for the routine analysis of samples, while the dark field technique is used to examine small metal particles on a support matrix.

Interpreting the TEM images of mesoporous silicas is not as straightforward as one might think. This is because TEM only provides a 2-D image while these mesoporous silicas all have a 3-D structure. For example, Fig. 2.6 shows two TEM images of an SBA-15 sample, which has a hexagonal long-range ordered structure. However, depending on the orientation of the sample particles, two very different images can be observed for the same sample. The image showing a hexagonal arrangement of pores was taken from a particle standing vertically on the sample grid. On the other hand, the image showing parallel structures was taken from a particle lying on the grid, as shown in the illustration. Other 3-D mesoporous structures, such as SBA-16 and MCM-48, are even more difficult to interpret. As a result, inexperienced researchers can misinterpret the result and think it might be a different structure. If the long-range order is observed, electron diffraction can be recorded using TEM (with higher operation voltage). Electron diffraction can help to confirm the structure of the sample.

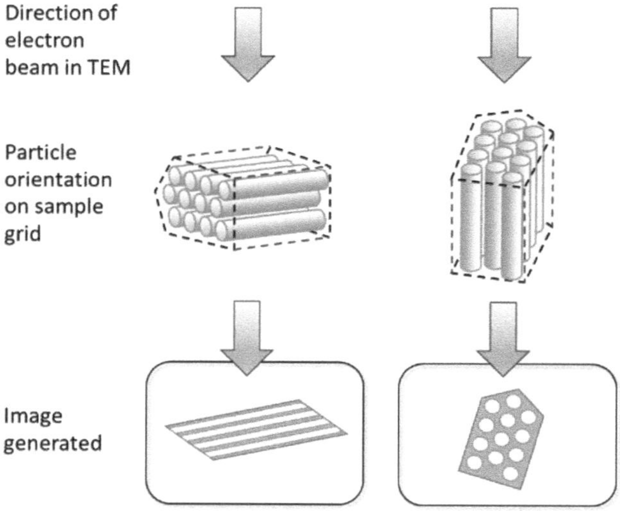

Figure 2.6 Illustration showing two different images can be generated from the same (e.g., SBA-15) sample due to different orientations.

2.3.2.2 Powder x-ray diffraction

In Chapter 1, we introduced powder XRD for characterizing nanoparticles. Powder XRD is also a common method to study the structure of mesoporous silicas. However, their diffraction patterns are not generated according to their atomic arrangement, or lattice structure, because mesoporous silicas are essentially amorphous. On the other hand, their regular porous structure is shown by the presence of characteristic reflections in the XRD pattern. For example, a hexagonal SBA-15 sample with a high degree of order shows a very sharp and intense [100] reflection. Minor reflections [110], [200], and [210] are also distinctive. A decrease in structural order is shown by the broadening of the [100] reflection while minor reflections become less distinctive and quite often disappear.

From the diffraction pattern of a sample, we can calculate the structural parameters, such as unit cells and d-spacing (or the distance between two repeated units), of a material. Once the d-spacing of the [100] reflection or d_{100}, is found, the pore dimensions can be estimated. The dimensions of the hexagonal unit cell, $a = 2d_{100}/\sqrt{3}$, can also be used for comparison between samples.

Let us have a recap. X-ray diffraction is mostly governed by the Bragg Law below,

$$n\lambda = 2d \times \sin\theta, \tag{2.1}$$

where n is the diffraction order (n = 1, 2, 3...) and θ is the glancing angle (the angle when the x-ray beam hit the sample). A diffraction pattern is usually plotted as the intensity of a diffracted beam in "counts" against 2θ. As the Bragg Law suggests, an increase in d-spacing leads to a decrease in the value of 2θ. For most powder XRD analysis of a crystalline sample, a standard scan of 2θ = 5 to 90° will suffice. For mesoporous silicas, such a scan is of no use as the materials are essentially amorphous at an atomic level, and there is no observable feature at $2\theta > 10°$ in general except the wide amorphous silica peak. Instead, mesoporous silicas require what we called a "low angle" scan, from 0.5 to 10°.

This caused much difficulty in the early years as many diffractometers manufactured in the 1980s and early 1990s cannot scan lower than 2° because of the danger of the incident x-ray beam directly hitting the detector without diffraction.

This could cause some considerable damage to the machine. Therefore, in some of the research papers from the earlier development of mesoporous silicas (e.g., pre-2000), the first [100] peaks were missing in XRD due to the limitations of the older diffractometers. This is particularly apparent for SBA-15 where the 100 peak is usually at $2\theta < 1°$. The advancement of modern diffractometers makes this possible and now many diffractometers are capable to scan from as low as 0.5°.

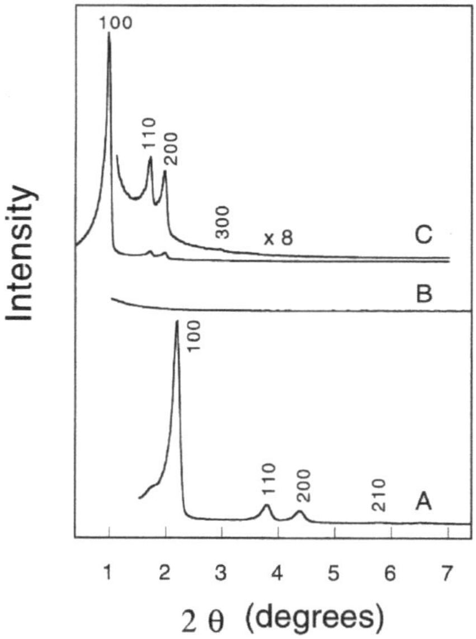

Figure 2.7 XRD patterns for (A) MCM-41, (B) MCM-41 with collapsed structure after being boiled in hot water for 6 h, and (C) SBA-15 (Zhao et al., 1998b).

Again, unlike crystalline materials, the diffraction patterns of mesoporous silicas are relatively simple and usually consist of only one main peak and two or three minor peaks, which can be quite broad. Such a feature makes a precise analysis of the structure using XRD very difficult. Other structural analysis techniques, usually TEM, are required as a supplement. Figure 2.7 displays some typical diffraction patterns of mesoporous silicas. Nowadays, many research papers leave out powder XRD analysis on mesoporous silica samples because the structural parameters from the techniques do not provide the data required for many applications, in particular pore volume and pore dimensions. Instead, many publications showed TEM images of the mesoporous silica samples as evidence for a long-range order.

2.3.2.3 Nitrogen adsorption/desorption isotherm

The porous structure of mesoporous silicas makes them unique. If we want to fully utilize such unique materials, their porous property needs to be fully understood. To acquire these essential data, gas adsorption is the most commonly used technique. Conventionally, the surface area, the mesoporous volume and the pore size distribution of mesoporous materials are the parameters that can be measured using nitrogen adsorption and desorption isotherms at −196°C. Nitrogen is chosen because it is inert and rarely chemically adsorbs on the surface of a solid. Also, when compared with other inert gases such as argon or krypton, nitrogen is by far the most widely available adsorbent.

The theory behind gas adsorption isotherm is quite simple. However, modeling the adsorption process is not so straightforward, in particular for solids with a porous structure. Generally, when the surface of a clean sample is exposed to a fixed amount of gaseous molecules, a portion of the molecules will be adsorbed onto the surface until an equilibrium is reached. This phenomenon is governed by the Langmuir equation. This also causes a consequent drop in the gas pressure. After achieving equilibrium, the amount of gas adsorbed can be calculated from the decrease in pressure by applying the gas law. In this case, the volume of the gas must be known and the temperature must be constant (hence the term "*adsorption isotherms*").

The adsorption of a gas by a solid is due to the attractive forces acting between the solid and gas molecules. Depending on the nature of these forces, the adsorption of a gas by a solid can be classified into two categories: physisorption and chemisorption. The relation between the amount of gas adsorbed, n, by a solid and the relative pressure, P/P^0 (where P is the vapor pressure and P^0 is the saturation vapor pressure at temperature T) at a constant temperature, T, is termed the adsorption isotherm. The adsorption isotherm can be expressed as V_{ads} (volume of adsorbate, e.g., N_2, adsorbed) vs P/P^0.

Every adsorption system (the solid sample or the absorbent) has its own characteristic adsorption isotherm. However, most of the physical adsorption isotherms can be classified into five major types (Types I to V, illustrated in Fig. 2.8), originally proposed by Brunauer, Deming, Deming and Teller (BDDT) in 1940. Figure 2.9 also shows the relationship between the pore size of a nanoporous material and its adsorption isotherm type.

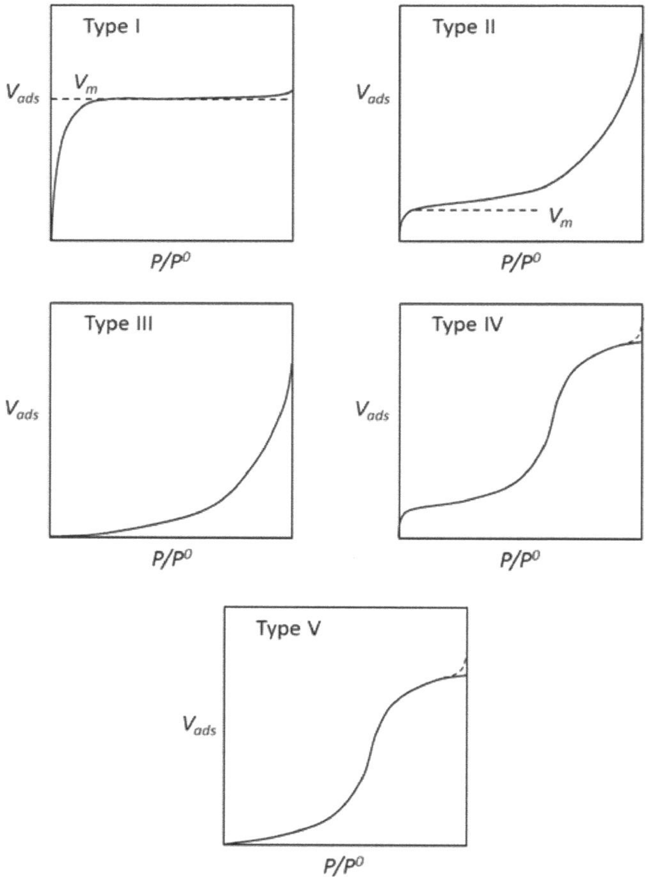

Figure 2.8 Type I to V adsorption isotherms. Generally, Type I: microporous materials; Type II: non-porous or macroporous materials; Type III: non-porous or macroporous materials with weak adsorbent-adsorbate interaction; Type IV: mesoporous materials; Type V: mesoporous materials with weak interaction. V_m indicates the monolayer capacity.

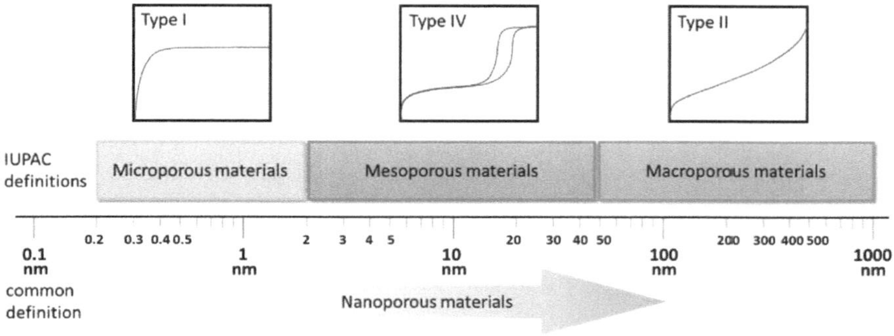

Figure 2.9 Relationship between pore size and adsorption isotherm type for nanoporous materials.

Mesoporous materials, with pore diameter of 2 to 20 nm, usually give rise to Type IV adsorption isotherms. A typical experimentally measured Type IV isotherm for an SBA-15 mesoporous silica sample is shown in Fig. 2.10. In the low-pressure region, a Type IV isotherm shows a similar path to a non-porous solid, i.e., Type II. However, the path deviates upward after a certain point C. This is due to capillary condensation of adsorptive gas inside the mesopores. It is very similar to the initial stage of a Type I isotherm for microporous materials.

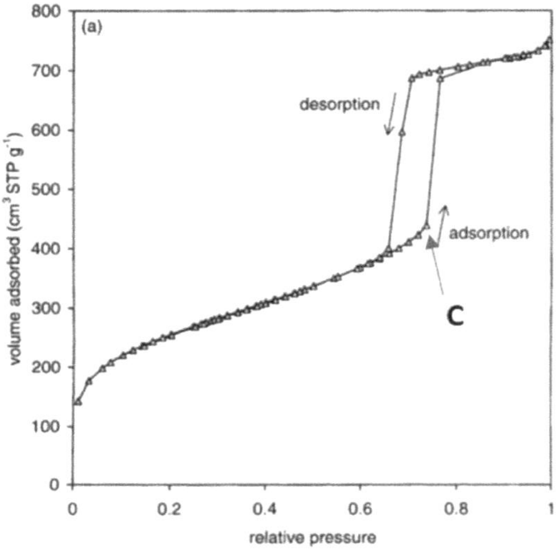

Figure 2.10 Nitrogen adsorption isotherm for SBA-15. (Yiu et al., 2001a).

At a higher pressure, the path reaches a plateau as the pores are filled. It then goes up again as saturation pressure is approached and condensation begins. From these isotherm data, we can calculate several key structural parameters (specific surface area, pore size distribution and pore volume) for a mesoporous material.

Box 2.3

Langmuir model based on the formation of the monolayer on a solid

The Langmuir model was developed for surface characterization. For a solid sample exposed to a given gas at a suitably low temperature, the gas will be physically adsorbed on the surface of the solid due to the interaction between the gas molecules and the surface of the solid. For a non-porous solid, adsorption increases with increased pressure of gas until monolayer coverage is achieved. The specific surface area A, i.e., surface area per 1 g of solid, can then be defined as

$$A = n_m a_m L \tag{2.2}$$

where n_m is the number of moles of gas adsorbed in the monolayer; a_m is the average area occupied by a gas molecule in the complete monolayer and L is the Avogadro's constant.

If θ_1 is the fraction of surface sites occupied by adsorbate molecules (the gas) and the fraction of the unoccupied sites is θ_0, the rate of condensation from the gas phase on a unit surface area becomes $a_1 \kappa p \theta_0$, where a_1 is the condensation coefficient, κ is the constant given by the kinetic theory of gases and p is the vapor pressure of the absorbate gas. On the other hand, the rate of evaporation from a unit surface area is expressed as $z_m \theta_1 v_1 e^{-q_1/RT}$, where z_m is the number of sites per unit area, v_1 is the oscillation frequency of the gas molecule and q_1 is the isosteric heat of adsorption. Therefore, at equilibrium, the rate of condensation and evaporation will be the same, hence,

$$a_1 \kappa p \theta_0 = z_m \theta_1 v_1 e^{-q_1/RT}, \tag{2.3}$$

when θ_0 is substituted by $1 - \theta_1$, this equation becomes

$$\theta_1 = \frac{a_1 \kappa p}{a_1 \kappa p + z_m v_1 e^{-q_1/RT}} \tag{2.4}$$

The Langmuir equation is derived when θ_1 is substituted by n/n_m where n is the number of moles of gas molecules adsorbed on the surface, and n_m is the number of moles of gas adsorbed in a monolayer,

$$\frac{n}{n_m} = \frac{Bp}{1+Bp}, \tag{2.5}$$

where $B = \frac{a_1 \kappa}{z_m v_1} e^{q_1/RT}$.

More conventionally, we may see equation 2.5 expressed in volume terms (V) as

$$\frac{V}{V_m} = \frac{K_{\text{eq}}^A p_A}{1+K_{\text{eq}}^A p_A}, \tag{2.6}$$

where A refers to the adsorbate (gas) and K_{eq} is the equilibrium constant of adsorption for A; A(g) ⇆ A(ads).

Box 2.4

BET model based on physical adsorption isotherm of a non-porous solid

In most cases, the assumptions made in deriving the Langmuir equation are not totally valid. In particular, coverage beyond monolayer is common. The Langmuir equation can be extended to multi-layer adsorption. In 1938, Brunauer, Emmett and Teller (BET) applied the Langmuir equation to multi-layer adsorption by introducing several simplifying assumptions. For condensation-evaporation equilibrium in the second adsorbed layer, the equation becomes,

$$a_2 \kappa p \theta_1 = z_m \theta_2 v_2 e^{-q_2/RT} \tag{2.7}$$

And for the ith layer,

$$a_i \kappa p \theta_{i-1} = z_m \theta_i v_i e^{-q_i/RT} \tag{2.8}$$

The fraction of surface sites occupied by i molecules becomes θ_i and the total number of gas molecules, Z, becomes

$$Z = Az_m(\theta_1 + 2\theta_2 + \ldots + i\theta_i) \tag{2.9}$$

Also, the amount of gas molecules adsorbed in moles is given by

$$n = \frac{Az_m}{L}\sum_{1}^{i}(i\theta_i) \tag{2.10}$$

Brunauer, Emmett and Teller introduced assumptions to simplify the above summation:

(1) The heat of adsorption of all layers, except the first layer, is equal to the molar heat of condensation, q_L.

(2) With the exception of the first layer, the condensation-evaporation conditions are the same in all layers, or $v_2 = v_3 = \ldots = v_i$ and $a_2 = a_3 = \ldots = a_i$.

The summation may then be expressed as

$$\frac{n}{n_m} = \frac{c\left(\frac{p}{p^0}\right)}{\left(1-\frac{p}{p^0}\right)\left[1+\frac{(c-1)p}{p^0}\right]}, \tag{2.11}$$

where $c = e^{(q1 - qL)/RT}$. This BET theory has been universally adopted and the equation become known as the well-known **BET equation**. After rearrangement for plotting, it becomes

$$\frac{\left(\frac{p}{p^0}\right)}{n\left[1-\left(\frac{p}{p^0}\right)\right]} = \frac{1}{n_m c} + \frac{c-1}{n_m c}\left(\frac{p}{p^0}\right) \tag{2.12}$$

The validity of the equation is indicated by plotting $\left(\frac{p}{p^0}\right)\Big/n\left(1-\frac{p}{p^0}\right)$ against $\left(\frac{p}{p^0}\right)$, which should give a straight line. This is termed by the BET plot.

The values of n_m and c can be found from the slope, s, and the y-intercept, i of the plot.

$$n_m = \frac{1}{s+i} \tag{2.13}$$

$$c = \frac{s}{i} + 1 \quad (2.14)$$

The specific surface area A of the solid sample can be calculated from n_m using equation. BET plots are usually valid between $0.05 < \frac{p}{p^0} < 0.3$. When $\frac{p}{p^0} > 0.3$, BET plots deviate from a straight line.

In mesoporous solids, the monolayer formation occurs in the same way as a non-porous solid because the size of pores is large enough for several layers of adsorbate to build up. Therefore, the front part ($0.05 < p/p^0 < 0.3$), or pre-hysteresis region, of a Type IV adsorption isotherm is almost identical to a Type II isotherm. The surface area of a mesoporous solid can be determined by the BET method. The resultant surface area figures are usually referred to as BET surface areas. We have to be aware that hysteresis may appear before $p/p^0 = 0.3$ because the condensation occurs at a lower pressure when the pore size gets smaller. For example, MCM-48 has a pore size around 2–2.5 nm and the BET surface area has to be calculated from the region of $0.05 < p/p^0 < 0.25$. Otherwise the surface area of the material will be over-estimated, since the BET plot deviates from a straight line. That is why sometimes we see an unusually high surface area (>1400 m^2g^{-1}) reported for MCM-48 materials. This problem is even more common for materials such as SBA-1 and other microporous materials including large-pore zeolites and metal-organic frameworks (MOFs). In fact, the BET theory is not suitable for microporous and many mesoporous systems with smaller pores, for which the Langmuir model is a better estimation for specific surface area.

Data from nitrogen adsorption/desorption isotherms can also be used to estimate the pore size distribution and the pore volume of the mesoporous sample. Such estimation is based on the theory of capillary action of a liquid, using the famous BJH model.

Box 2.5

Pore size distribution (PSD) estimated using the BJH model

The most popular method for estimating the pore size distribution of a mesoporous material is the **Barrett, Joyner and Halenda** (BJH) method, which applies the Kelvin equation to correlate the radius of pores with the relative pressure of nitrogen in equilibrium during capillary condensation. This method is only applicable for Type IV isotherms and a cylindrical pore model is assumed, which includes open-ended pores and in the extreme, no pore networks.

Starting from the Young-Laplace equation, which relates the differential pressure, Δp, across the gas/liquid meniscus inside a pore to the radii of curvature (r_1 and r_2, see Fig. 2.11 below) and the surface tension of liquid, γ, the Kelvin equation can be derived.

$$\Delta p = \gamma\left(\frac{1}{r_1} + \frac{1}{r_2}\right) \tag{2.15}$$

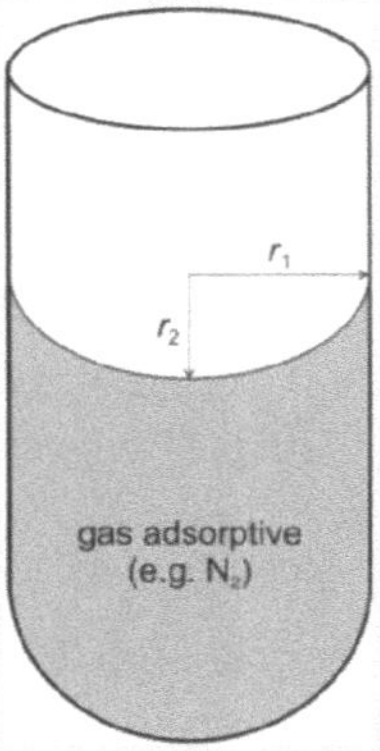

Figure 2.11 Illustration of a liquid meniscus in a pore.

Alternatively, for convenience in pore size determination, the Kelvin equation can be rewritten in terms of r_m, which is the mean radius of curvature of the meniscus by substituting $\frac{1}{r_m} = 2\left(\frac{1}{r_1}\right) + \left(\frac{1}{r_2}\right)$,

$$\Delta p = \frac{2\gamma}{r_m} \tag{2.16}$$

At a constant temperature, the Kelvin equation can also be written by relating the partial pressure p/p^0 to the mean radius of curvature, r_m, of the meniscus,

$$\ln\frac{p}{p^0} = -\frac{2\gamma V_L}{RT}\left(\frac{1}{r_m}\right), \tag{2.17}$$

where V_L is the molar volume of the gas adsorbate and γ is the surface tension of the liquid adsorbate (after condensation).

It is important to note that condensation occurs after the pore walls are completely covered by a layer of thin adsorptive film of thickness, t. Therefore, the pore radius (r_p) is

$$r_p = r_m + t \tag{2.18}$$

Substituting (2.17) into equation (2.18),

$$r_p = \frac{2\gamma V_L}{RT \times \ln\frac{p^0}{p}} + t \tag{2.19}$$

It is possible to convert r_m into a pore size if the pore shape and the angle of contact, ϕ, between the capillary condensate and the adsorbed film on the walls are known.

In the determination of the pore size distribution using the Kelvin equation, the desorption isotherm is normally used. This is because the Kelvin equation is based on the liquid-vapor equilibrium of a fully filled pore system. This can be explained in terms of thermodynamics. The Kelvin equation can be derived on a thermodynamic basis. Assuming dn moles of vapor in equilibrium with the bulk liquid at pressure P_0 is transferred into a pore where the equilibrium pressure is P. The free energy change is only due to the expansion of the vapor from P_0 to P and therefore

$$dG = \left(RT\ln\frac{P}{P_0}\right)dn, \tag{2.20}$$

where dG is the free energy change.

For a typical Type IV adsorption-desorption isotherm (Fig. 2.9) the desorption isotherm is always above the adsorption isotherm in the hysteresis region. Considering the thermodynamics in this region, the molar free energy change due to the condensation of the vapor into a pore during adsorption is

$$\Delta G_{\text{ads}} = \ln P_{\text{ads}} - \ln P_0 \tag{2.21}$$

and that corresponding to the desorption isotherm is

$$\Delta G_{\text{des}} = \ln P_{\text{des}} - \ln P_0 \tag{2.22}$$

From the adsorption-desorption isotherm, at the same quantity of adsorbed adsorptive, $P_{\text{des}} < P_{\text{ads}}$, and thus $\Delta G_{\text{des}} < \Delta G_{\text{ads}}$. Therefore, the desorption value of relative pressure corresponds to the more stable adsorbate condition and the desorption isotherm should be used for pore size analysis.

The use of the Kelvin equation is also limited to mesoporous materials. When the pore diameter is smaller, the values of V_L and γ of equation 2.19 deviate from that of a bulk liquid system and the Kelvin equation is no longer valid because the isotherm will be in a transition towards Type I (microporous materials). The limit in the pore diameter for the application of the Kelvin equation is usually about 2 nm. Therefore, this method of pore size estimation is not applicable to zeolites and metal-organic frameworks (MOFs), which typically have pore diameters less than 2 nm.

2.3.2.4 Silicon solid-state NMR

In the beginning when scientists had very little understanding of mesoporous silica materials, many techniques available were used for their characterization, including solid-state NMR, in order to understand these new materials. Magic angle spinning solid-state NMR (or MAS NMR) provides information about the coordination of atoms of Si and other NMR active elements in the lattice. As mentioned previously, although mesoporous silicas exhibit long-range order in their structures, they are essential non-crystalline. As a result, diffraction techniques including powder XRD and neutron diffractions can only provide simple structural information such as the repeating distances (or d-spacing) between pores. With the use of solid-state NMR, which is sensitive to short-range order, the local environment of Si atoms can be studied.

When MCM-41 was first reported in 1992, ^{29}Si MAS NMR was used to study the environment of Si atoms in the lattice walls (Beck et al., 1992). In MCM-41 the silica walls have a finite thickness and ^{29}Si MAS NMR proves particularly useful for distinguishing between Si atoms in the bulk of the silica walls and those on the surface of the wall. Such analysis can be carried out for all mesoporous silica materials including SBA-15 and MCM-48.

In a silica material, the basic structural units of the framework are SiO_4 tetrahedra, which link together by sharing oxygen atoms. For simplification, these SiO_4 units can be classified into Q^n units. Generally, Q^4 corresponds to the three-dimensional silica framework **Si**$(OSi)_4$ units and Q^3 is the surface **Si**$(OSi)_3O^-$ unit. From the Q^4: Q^3 ratio observed from the NMR spectrum, one can estimate the thickness of the silica wall of a mesoporous silica sample. Unfortunately, in the early years, i.e., 1990s, solid-state NMR spectra were always noisy and of low resolution, ^{29}Si MAS NMR is no exception. Therefore, deconvolution of the recorded spectra is usually required for the calculation of the Q^4:Q^3 ratio. Also, the natural abundance of ^{29}Si is only 4.7% and it has a long relaxation time. This leads to a long experimental time for one measurement.

Nowadays, the structure of mesoporous silica materials is well-understood. Fewer and fewer research papers report on the NMR analysis of mesoporous silicas. This is also partly due to the long experimental time and the unavailability of a solid-state NMR spectrometer in many research institutes.

2.4 Applications of Ordered Mesoporous Silicas

Thanks to their unique porous structures and versatility in terms of surface functionalization, ordered mesoporous silicas have numerous applications ranging from biomedicine to electronics. Here we introduce a few examples which are likely to have some impact in scientific areas.

2.4.1 Use as Templates for Making Other Nanostructures

The unique ordered structures of mesoporous silica provide great templates for preparing other ordered porous materials that are not possible to be made via liquid crystal templating or other methods. These materials include carbon-based CMK materials (which will be burnt during the calcination step) and other inorganic materials that have no stable inorganic precursors. The synthesis and applications of these materials will be discussed in detail in Chapter 3.

2.4.2 Separation and Chromatographic Applications

Chromatography is a category of separation techniques based on the interaction between a "stationary phase" and the elute (molecules that are to be separated). If the surface chemistry of the molecules in the elute is different (e.g., hydrophilicity vs hydrophobicity), they will have different interactions with the support. Providing a large enough overall surface area on the stationary phase for the elute to interact, molecules can be separated. Figure 2.12 illustrates this phenomenon.

Silica was widely used as the stationary phase for chromatography when it was first developed. Chromatography-grade silica gel has a high surface area (>200 m^2/g) but ordered mesoporous silica can be much better than commercial silica gel in terms of surface area (5-fold improvement). Therefore, ordered mesoporous silicas have been used as a stationary phase for both gas chromatography (GC) and high-performance liquid chromatography (HPLC) (e.g., Grün et al., 1996; Zhao et al., 2002).

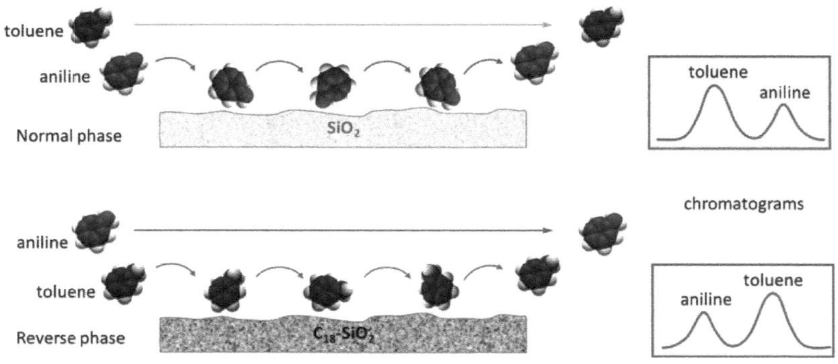

Figure 2.12 Illustration showing the separation of toluene from aniline on a standard and C_{18}-functionalized SiO_2 surface.

Similar to silica gel, mesoporous silica can also be functionalized via silanization (or post-synthesis grafting) to convert its surface into a hydrophobic surface using alkyl-silane (e.g., C18, *n*-octadecyltriethoxysilane) or phenyl-silane (e.g., phenyltriethoxysilane). Due to the much higher specific surface area, the separation between polar from non-polar molecules can be enhanced. In addition to a hydrophobic surface, silanization allows other functionalities to be grafted onto the silica surface. Methods (*in situ* and post-synthesis) for grafting these functional groups using silanization have been discussed earlier in Section 2.2.4. The most commonly used ones are grafting amine ($-NH_2$) groups and sulfonate ($-SO_3H$) groups for positively and negatively charged surfaces.

In the literature, organically functionalized mesoporous silicas (or PMOs) have been used as a stationary phase for GC and HPLC applications. For example, benzene (phenyl) and diethylbenzene-bridged PMOs were used as the GC stationary phase to separate substituted aromatic compounds (Borghard et al., 2009). But more interestingly, mesoporous silicas have been functionalized to carry a chiral surface for the separation of chiral compounds. For instance, He et al. demonstrated using (R)-(−)-phenylglycinol as the chiral agent on a sodium dodecylsulfate (SDS)-directed mesoporous silica as the chiral stationary phase for the separation of several racemic compounds including citronellal and limonene (He et al., 2019). Regarding HPLC, one early example was published in 2006 when phenylene-bridged PMO spheres were used as the HPLC stationary phase to separate aromatic compounds (e.g., benzene, naphthalene, biphenyl and phenanthrene) (Rebbin et al., 2006). Chiral HPLC separation has also been demonstrated using trans-(1R,2R)-diaminocyclohexane-functionalized PMOs as Zhu et al. used this system to separate chiral amino acids (Zhu et al., 2007). One key issue for use in HPLC for the stationary phase is to overcome back pressure. The mechanical strength of these PMOs and, more crucially, their spherical morphology is the solution.

Due to the ordered porous structure of functionalized mesoporous silica or PMOs, their advantages over other conventional stationary phase materials (e.g., silica gel) were shown in the separation of biomolecules. The narrow pore size distribution (PSD) provides an excellent "selective binding" control according to the molecular size of biomolecules in the eluent. An early example has shown that thiolated SBA-15 has a distinctive cut-off point for around 35 kDa (see Fig. 2.13), i.e., size exclusion. (Yiu et al., 2001b). The thiolated surface has been shown to have effective interaction with many protein molecules (partly due to the cysteine residue in structure). This work showed that in addition to the size exclusion effect, the thiolated surface induced a reversible binding and little blockage was shown even with relatively large protein molecules such as β-lactoglobulin.

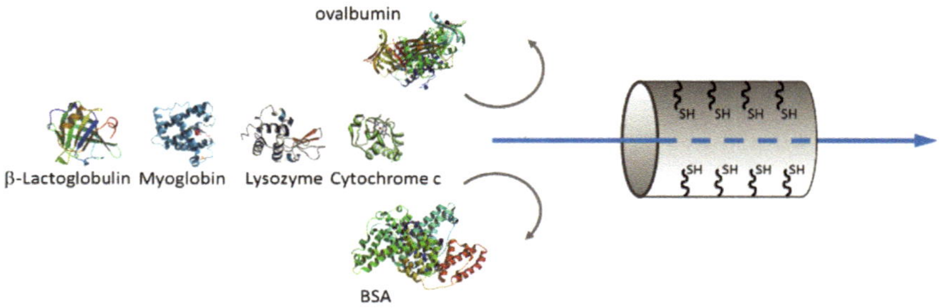

Figure 2.13 Illustration showing the size exclusion effect of mesoporous silica on protein molecules.

Another surface-biomolecular interaction that has been studied was the electrostatic interaction as well as the size exclusion effect of mesoporous silica. For example, Han et al. studied the adsorption behavior of three negatively charged proteins conalbumin, (77 kDa, pI = 6.0), ovalbumin (44 kDa, pI = 4.9) and trypsin inhibitor protein (14 kDa, pI = 5.1) onto APTES-grafted SBA-15 and MCF (an extended materials that will be introduced in Chapter 3). A clear preferential adsorption on the trypsin inhibitor was shown by the APTES-grafted SBA-15 sample (Han et al., 1999).

Recently, concern about our environment has been a growing topic and materials for effective removal of pollutants is an expanding area of research. With the high surface area of functionalized mesoporous silica, there is a good opportunity to develop excellent adsorbents from its variety of surface properties. At the early stages of research on functionalized mesoporous silica, scientists have worked on using functionalized mesoporous silica to remove pollutants in aqueous media (e.g., freshwater sources). For example, thiolated MCM-41 has been used to adsorb heavy metal ions including Cu^{2+}, Cr^{2+}, Pb^{2+} and Ag^{+} (Wu et al., 2010). Interestingly, it was also found that the counter anion had a certain effect on the adsorption capacity of Cu^{2+}, with Cl^{-} showing the highest adsorption while NO_3^{-} having the lowest. Another functional group that can be used for the adsorption of heavy metal ions is the amino group $-NH_2$, which has the capability to act as a ligand and form a transition metal-nitrogen coordination bond. For instance, aminated MCM-41 has been incorporated into a Nylon-6 membrane for removing Cu^{2+} and Cd^{2+} ions from spiked water samples (Mohammadnezhad et al., 2017). The problem with these systems lies in the cost of the production of functionalized mesoporous silica and their availability in large quantities (e.g., in tons). Environmental uses of adsorbents are usually in very large quantities; so cheap, widely available but less sophisticated adsorbents (e.g., activated charcoal, zeolites and ion-exchange resins) are still more popular for the removal of heavy metals.

2.4.3 Catalysis

2.4.3.1 Solid acid catalysts

As mentioned at the beginning, these ordered mesoporous silica materials were originally thought to be used as large-pore zeolites. The increased pore dimensions are likely to minimize the problem of coking and blockage experienced from using zeolites in catalytic reactions. Due to the small pore size, the formation of coke from trapped organics in zeolites will badly deactivate the catalysts by blocking the active sites inside the pores. This problem is less likely to happen if ordered mesoporous silica is used because the large pores are unlikely to be blocked by coke as easily as in the case of zeolites. Mesoporous aluminosilicate (Al-MCM-41) was the first ordered mesoporous catalyst reported in the literature (Kresge et al., 1992). The incorporation of aluminum into the silica framework is to create Si–OH···Al sites, which are Brønsted acid sites, commonly found in aluminosilicate zeolites.

In zeolite systems, aluminum substitution on a silica framework generates Brønsted acid sites due to the charge balance between Al and Si atoms (see Fig. 2.14). This can also be achieved in mesoporous silicas. Indeed, in the Mobil report in 1992, Al-MCM-41 was already reported and tested as a solid acid catalyst (Kresge et al., 1992). However, the strength in acidity was affected by the low crystallinity of the SiO_2 framework in MCM-41 (also seen in other mesoporous silicas). Moreover, the amount of Al substitution (the reverse of the Si/Al ratio) is also limited as increasing Al substitutions cause structural collapse. Consequently, the synthesis and application of Al-substituted mesoporous silicas have not been widely reported since 2000. Instead of Al, other elements with a +3 valency (e.g., B, Fe or Ga) can also have the same effect in generating acid sites. Table 2.3 shows some examples of Al-substituted mesoporous silica for catalytic applications.

Similar to zeolite too, these acid sites generated from Al substitution can also behave as cation exchange sites (see Fig. 2.14). Again, due to the same reason as its solid acid properties, the number of cation exchange sites in the framework of mesoporous silicas can be very limited. Further, cations of interest (e.g., metal ions) are generally small enough to travel through the micropores of zeolites, there is little interest to develop a mesoporous cation exchanger. Hence, this area has not been fully developed.

There are two methods for Al substitution. Aluminum precursors (e.g., aluminum alkoxides) can be added to the reaction mixture in tandem with the silica source during mesoporous silica synthesis (one-pot synthesis). In a different pathway, aluminum precursors can be adsorbed onto the surface of freshly calcined mesoporous silica, followed by thermal decomposition (post-synthesis). The first method tends to be more commonly used since Al leaching has been reported from materials prepared with the second method.

If substitution on a mesoporous silica framework can generate acid sites and cation exchange sites, there must be an element capable to generate basic sites and anion exchange sites via substitution. Because these sites are generated by charge balance, the element needs to have an oxidation of +5, which is not very common. One successful example of generating basic sites on mesoporous silica is niobium Nb (Zhang and Ying, 1997). Upon Nb-substitution, MCM-41 can behave as a solid base and should be active in basic catalysis. However, most research on Nb-substituted mesoporous silica is either on acid- or on redox catalysis. One problem with developing solid base catalysts is that solid bases are widely available globally, unlike solid acids. For example, quick lime CaO is a good solid base material and it is cheap. There is no significant market value for niche solid base catalysts.

Table 2.3 Early examples of substituted mesoporous silica and their applications

Element	Mesoporous silica type	Si:M ratio	Acid/ base	Example reactions	Reference
Al	Al-MCM-41	95	Acid	Friedel–Crafts alkylation	Armengol et al., 1995
	Al-MCM-41 (Na^+, Cs^+ exchanged)	13	Acid	Aldol condensation	Kloetstra & van Bekkum, 1995
			Base	Knoevenagel condensation	
	Al-FSM-16	20	Acid	Synthesis of tetraarylporphyrins	Shinoda et al., 1995
B	B-HMS	7–166	Acid	**Not tested*	Tuel & Gontier, 1996
	B-MCM-41	5-40	Acid	**Not tested*	Liu et al., 1996
Fe	Fe-MCM-41	100–200	Acid	Oligomerization of propene	Kim & Inui, 1996
	Fe-HMS	15–85	Acid	**Not tested*	Tuel & Gontier, 1996
Ga	Ga-MCM-41	100–200	Acid	Oligomerization of propene	Kim & Inui, 1996
	Ga-MCM-41	25–71	Acid	**Not tested*	Cheng & Klinowski, 1996
	Ga-MCM-41	10–120	Acid	**Not tested*	Cheng et al., 1996
	Ga-HMS	13–86	Acid	**Not tested*	Tuel & Gontier, 1996
Nb	Nb-MCM-41	10–100	Base	**Not tested*	Zhang & Ying, 1997

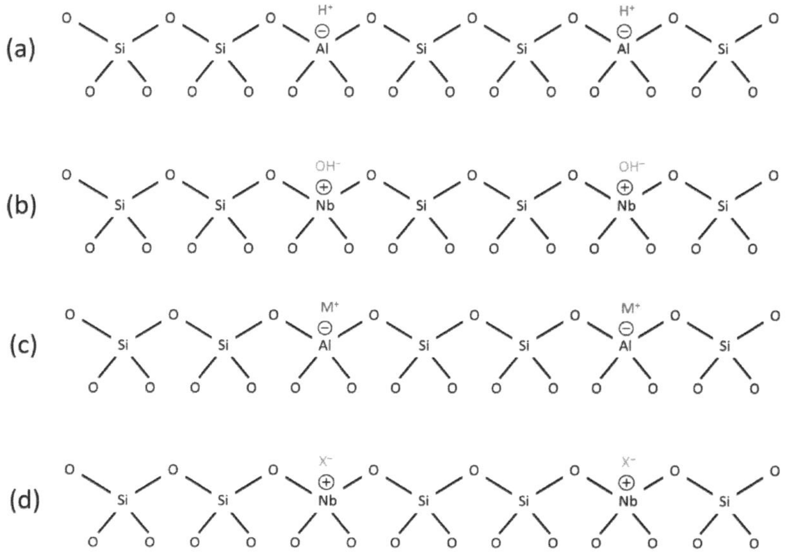

Figure 2.14 Illustration of the acid (a) and basic (b) sites of substituted silica. (c) and (d) also illustrate the nature of the cation and anion exchange sites.

Although these mesoporous aluminosilicate materials showed some acid catalytic activities, the recorded activities were always much lower than that of zeolites, such as ZSM-5 or zeolite-Y. This is believed to be linked to their poor crystallinity. Scientists then turned to a group of strong acid catalysts called "superacids," which are species of a "Hammett acidity function, H_0" (a scale for measuring acidity) < 0. For example, hydrogen fluoride (HF) has an $H_0 = -15.1$ while pure sulfuric acid (H_2SO_4) has an $H_0 = -11.9$. However, not all superacids are suitable to be supported on mesoporous silica due to the severe problem of leaching from support. Both H_2SO_4 and HF are likely to leach out due to their small size and hence would not be good candidates, no matter how strong they are. In fact, HF is likely to dissolve the mesoporous silica support too. There is a group of superacids considered to be suitable for such application. Heteropoly acids, or HPA, are anionic clusters with metal oxide tetrahedrals (MO_4) structures. For example, $H_3PW_{12}O_{40}$, a tungstenphosphonic acid, is an HPA with 12 WO_4 clusters in a famous "Keggin structure." These HPAs are small enough to fit inside the mesopores of mesoporous silicas but retain their high acidity. $H_3PW_{12}O_{40}$ has been supported inside MCM-41 as a solid acid for a trans-de-*t*-butylation reaction (Kloetstra et al., 1996; Kozhevnikov et al., 1996). This substrate was chosen possibly because it is bulky in size and unlikely to go through a zeolite micropore. Despite some early success, research on supported HPA on mesoporous silica did not take off after some early efforts in the late 1990s.

The surface of mesoporous silica can also be made acidic by grafting organic acid groups. As shown in the POMs section earlier, nitrile groups (–C≡N) can be grafted onto the silica wall via the post-synthesis silanization route. This nitrile group can undergo further reaction to form carboxylic acid when it is heated in a mild acid (e.g., 1 M HCl, see Fig. 2.14 below). However, carboxylic acid is a weak acid and it is unlikely to have a high value in catalytic applications (although it can be useful in other areas of application, including biomedicine and separation). On the other hand, thiolated mesoporous silica (prepared by grafting mercaptopropyltri-ethoxy silane) can be oxidized using hydrogen peroxide (H_2O_2) or nitric acid (HNO_3) to form a sulfonic acid group, which is a much stronger acid than carboxylic acid (see Fig. 2.15). For example, van Rhijn et al. have demonstrated using sulfonic acid-functionalized MCM-41 for

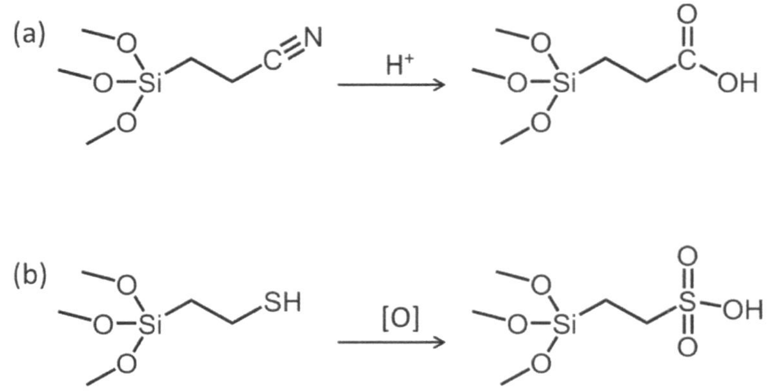

Figure 2.15 Schematic showing two pathways to generate acidic groups via silanization: (a) acidification of a nitrile group and (b) oxidation of a thiol group.

condensation, esterification and oligomerization of furan (van Rhijn et al., 1998a; van Rhijn et al., 1998b). This direction of research has attracted more attention than that of supported HPA. There are still research papers published in 2019 reporting the use of sulfonic acid-functionalized mesoporous silica as solid acid catalysts (Doustkhah et al., 2019).

2.4.3.2 Redox catalysts

In order to generate mesoporous redox catalysts, several transition metals have also been incorporated into mesoporous silica frameworks. One early example was the incorporation of titanium (Ti) into the MCM-41 structure for oxidizing reaction, oxidation of 1-hexene with H_2O_2, reported by Blasco et al. (Blasco et al., 1995). The aim of this work is to develop a new "large-pore" oxidative catalyst suitable for larger substrates. Although this Ti-MCM-41 showed good selectivity towards the desired epoxide product, the record activity was much lower than two selected Ti-zeolites (TS-1 and Ti-Beta). Nonetheless, it showed improved activity in non-aqueous media since the surface of Ti-MCM-41 was found hydrophobic.

Table 2.4 Some early examples of substituted mesoporous silica aimed for redox applications

Element	Mesoporous silica type	Si:M ratio	Example reactions	Reference
Ti	Ti-MCM-41, Ti-MCM48, Ti-KIT-1, Ti-SBA-1, Ti-HMS	49	Oxidation of phenol and 2,6-DTBP	Kang et al., 2000
	Ti-MCM-41	100	Oxidation of alkenes and alkanes	Tatsumi et al., 1998
	Ti-MCM-48	50	Oxidation of alkenes and alkanes	Tatsumi et al., 1998
	Ti-HMS	10–100	Oxidation of phenol and 2,6-DTBP	Jang et al., 2005
V	V-HMS	30–120	**Not tested*	Gontier & Tuel, 1995
	V-MCM48		Oxidation of phenol and 2,6-DTBP	Kang & Ahn, 2000
Sn	Sn-MCM-41	83–178	Hydroxylation of phenol and 1-naphthol	Das et al., 1995

Vanadium-substituted mesoporous silica is another possible candidate catalyst for oxidation reactions being studied in the early years. Reddy et al. reported the synthesis of V-MCM-41 for the oxidation of cyclododecane and 1-naphthol with H_2O_2 as an oxidant (Reddy et al., 1994). These two bulky organic substrates were selected to show that the mesoporous structure of MCM-41 was suitable for larger organics, which microporous zeolites would not be able to accommodate. However, in both cases, the recorded conversions were relatively low (34% and 11%, respectively). This result was not so promising. Table 2.4 shows selected examples of metal-substituted mesoporous silicas with their catalytic reactions.

2.4.3.3 Metal nanoparticles supported on mesoporous silicas

The catalytic properties of metal nanoparticles, including Pt, Pd, Ni, have been well studied for decades. To enhance their catalytic performance, these nanoparticles are commonly attached onto a support material, usually of a high surface area. You may find such catalysts in many daily life sources. Catalytic converters in modern cars are supported Pd. Supported metal particles on silicas are widely used catalysts in industries for many redox reactions and are commercially available. The high surface area and porous structure of mesoporous silicas make them good candidates as the support for these metal nanoparticle catalysts.

Generally, metal nanoparticle-supported catalysts can be made using wet impregnation method. The catalyst support, i.e., the mesoporous silica, is soaked in the metal precursor solution and dried for deposition. Usually a metal nitrate (e.g., $Ni(NO_3)_2$) is used as a precursor if available while the solvent can be a polar, volatile solvent such as ethanol. Metal oxide on mesoporous silicas can then be formed by thermal decomposition. In this form, the catalysts can be stable for storage for several months to a year. Reduction to the active form of metal nanoparticles on support needs to be carried out just before use for reaction. Using this preparation method, the particle size of these metal nanoparticles can be as small as 2–3 nm, hence a high metal surface area.

One of the first examples is a simple Pt on MCM-41 catalyst for CO oxidation (Junges et al., 1995). The catalyst was prepared with the wet impregnation method using Pt salts, $Pt(NH_4)_2](NO_3)_2$, and $K_2[PtCl_4]$, with up to 4% w/w loading. A 100% conversion was recorded at 80°C. A Pd/Nb-MCM-41 catalyst was then reported for a series of Heck reactions, prepared from chemical vapor deposition (CVD) (Mehnert and Ying, 1997). The average particle size of Pd was found to be 5 nm. The Nb on MCM-41 offered a basic surface that the Heck reaction requires. One other direction of this research is moving towards environmental applications and hydrogen purification. As Junges et al. demonstrated using Pt/MCM-41 catalyst, selective oxidation of CO can help to purify hydrogen from a syn gas product. Huang et al. published a comprehensive review in this area (Huang et al., 2009).

Another direction is to prepare bimetallic catalysts using mesoporous silicas as supports. In many reactions, bimetallic catalysts show improved activity over single-metal catalysts. It is also possible to prepare mesoporous silica-supported bimetallic catalysts using a mixed salt precursor (Thomas et al., 2003). Again, the recent trend points towards environmental research since these catalysts are capable to remediate/degrade organic pollutants under mild conditions (Quiton et al., 2021).

2.4.3.4 Transition metal complexes on nanoporous silicas

One advantage of mesoporous silicas that zeolites could not match is their ability to graft large organic molecules onto their surface via silanization. This allows the immobilization of a wide range of homogeneous catalysts. Homogeneous catalysts, including many transition metal complexes, are powerful catalysts and they are usually more active than heterogeneous catalysts due to enhanced mass transfer. However, their main issues are difficulties in recycling and many transition metal complexes are expensive. Moreover, they also prosses environmental issues since many of these transition metals (e.g., Co, Ni, Cu, Cr) are toxic. Immobilization on a support can minimize their environmental impact and allow recycling, making their use more economically viable.

Immobilizing transition metal complexes onto mesoporous silicas have two potential merits; (1) their high surface area can allow a high loading of catalyst, and (2) their narrow pore size distribution may add the further advantage of size selectivity. Plenty of examples have been reported in the literature. The first example was reported as early as 1995 when Brunel et al. presented a post-synthesis route in grafting various ligands on the MCM-41 surface (Brunel et al., 1995). Despite no test on catalytic activity, this early paper paved a path for future development in this area. For example, salicylaldimine ligands were grafted onto MCM-41 via silanization (Ray et al., 2007). Transition metals (Co, Ni and Cu) were coordinately bound onto the salicylaldimine ligands and the catalysts were tested with a model reaction, hydroxylation of phenol. When compared with the same complexes supported on amorphous silica, the MCM-supported catalysts favored the hydroquinone product over catechol. This was believed to be the effect of pore size favoring the formation of "narrower" hydroquinone molecules. In a related work, a similar strategy was employed to graft salen ligands onto SBA-15, followed by metal coordination with Co(II), Fe(III) and Cu(II). These catalysts were tested with the oxidation of 5-hydroxymethylfurfural using H_2O_2 as the oxidant and the supported Cu(II) salen showed the highest activity with 2,5-diformylfuran being the major product. However, no comparison test was carried out against amorphous silica supports. Remarkably, the supported Cu(II) salen catalysts retained over 96% of the original % conversion after 6 cycles of use, suggesting that the reuse of the catalyst is possible (Martínez-Vargas et al., 2017). In an earlier example, manganese(II) bipyridine $[Mn(bipy)_2]^{2+}$ complex was supported onto Al-MCM-48 through electrostatic interaction (Eswaramoorthy et al., 1998). The catalyst was tested using the oxidation of styrene to styrene oxide. Interestingly, the 3D porous network of Al-MCM-48 as a support outperforms the 2D hexagonal porous system of Al-MCM-41, despite having a large pore size. This is possible because 2D systems are more likely to experience blockage and slow down the diffusion of reactants and products. Such observation also reminds us that although mesoporous silicas have enlarged pore sizes, as seen in supported metal nanoparticle catalysts, the incorporation of a foreign species, such as a metal complex, will inevitably narrow down the pores and slow down diffusion.

2.4.4 Enzyme Immobilization

Enzymes are highly valuable catalysts because of their specificity and high activity in mild conditions. Moreover, their enantioselectivity is still not matched by most synthetic catalysts. However, the major setback of using enzymes on industrial scales lies in their high associated costs. It is because most enzymes are still extracted from biological species such as fungi, bacteria, and animals. The purification process is costly and the productivity through these biological sources is still low. Hence, scientists have been exploring possibilities for easy recovery, recycling and reuse of enzymes, in order to make their use commercially viable. Immobilizing enzymes on the surface of a solid carrier is a logical way because the catalysts can then be separated by filtration. Before the discovery of mesoporous silica, enzyme immobilization was carried out on bulk materials such as silica particles, polymer particles or even by crosslinking enzyme molecules with each other. All these immobilization methods work to a certain extent. However, crosslinking has the risk of denaturing the enzyme molecules. The use of silica or polymer particles, due to their low surface area, experienced the problem of low loading. New support materials were required to take enzyme immobilization to the next level.

Many high surface area materials, such as zeolites, are simply incompatible with enzymes in terms of dimensions. Zeolites are microporous materials with pores of diameter < 2 nm, while most enzymes are of a size > 5 nm. The vast internal space of a zeolite is simply wasted for enzyme immobilization. Despite this, research on using zeolites for enzyme immobilization can still be found in the literature (Taghizadeh, 2020; Lu et al., 2022). The logical development in this area will be the use of large-pore zeolites, which are rare.

Indeed, when mesoporous silicas were first discovered, scientists treated them as large-pore zeolites. However, because of their lack of success in out-performing the existing catalysts such as zeolites, e.g., ZSM-5 or fajusites (zeolite-Y), scientists started to look elsewhere for their use in catalysis. In 1996, Díaz and Balkus published the first paper using MCM-41 silica as a carrier for various enzymes, including cytochrome C, trypsin, papain, and horseradish peroxidase (Díaz and Balkus Jr., 1996). Two problems arose from this work; the pore size of MCM-41 is only 2.5–3 nm wide while many of these enzymes are too large to fit inside the pores. Secondly, the leaching of enzymes from the carrier is still a problem. Pore expansion by adding oil, such as mesitylene (trimethylbenzene), was suggested to be the solution for the first. However, the use of mesitylene causes the loss of long-range order and there is a limit of how much the pores can be expanded, usually < 4 nm without sacrificing the long-range order. The second was dealt with by closing the pore opening with 3-aminopropyltrimethoxysilane (APTMS). Though the problem of leaching was minimized, many enzyme molecules were denatured with this process due to the heat, the chemicals and the organic solvents used for pore closing.

Despite these setbacks, research in using mesoporous silica for enzyme immobilization started to roll. In 2001, Yiu et al. demonstrated the use of SBA-15 silica functionalized with various organic groups for enzyme immobilization (Yiu et al., 2001a). As such, the problems arising from Díaz and Balkus' work were solved. The increase in pore size using SBA-15 allows the immobilization of larger enzyme molecules, while the organic surface minimized the leaching problem without the post-immobilization sealing process. In this work, the trypsin immobilized on thiol-SBA-15 was found to be recyclable with 66% of activity retained. Since then there has been a significant development in this area (Carlsson et al., 2013). Many research works have since been published based on these early two papers.

2.4.5 Drug Storage and Delivery

The extra high pore volume, up to 1.2 cm^3/g, of ordered mesoporous silica materials is perfect to use as storage devices for many functional substances for delivery. One current example is the use of ordered mesoporous silica as drug carriers. There are a number of unique properties of these materials making them superior to other competitors; (1) their high pore volume leads to a high drug loading capacity; (2) the surface of these materials can easily be functionalized to enhance/control the drug-carrier surface interaction; (3) the high surface area (up to

1000 m^2/g) allows a high functional group concentration. This in turn further enhances the drug loading capacity; (4) these materials in general show low cytotoxicity; (5) these materials have been shown to be degradable in body simulated fluid (BSF). This suggested that the carriers can be breakdown by the human body and discharged from bodies after the delivery of drugs.

One of the first examples was demonstrated by the research group led by Vallet-Regi in Spain. In this work, ibuprofen was used as a model drug and SBA-15 was the testing material (Vallet-Regi et al., 2001). Ibuprofen is commonly used as a model drug because it is easily available and the analysis using UV-visible spectrophotometry is relatively simple. However, delivery of ibuprofen is unlikely to be of any significance because ibuprofen is only a simple anti-inflammatory drug. Nonetheless, the findings gave some important guidelines for scientists to understand the important criteria for a successful mesoporous system to deliver high-value drugs. Since then, many other drugs have been used in similar studies. Anti-cancer agents are one of the major categories of drugs experienced some major advancement (He et al., 2011). This is mainly because these drugs are usually expensive and they tend to cause severe side effects to healthy tissues. Targeted delivery of anti-cancer agents can both save costs and reduce unwanted side effects by localizing the drugs towards the tumors. For example, Slowing et al. demonstrated using folic acid-functionalized MCM-41 to target cancer cells since folate receptors are widely considered as cancer markers (Slowing et al., 2006). Other examples include using triphenoylphosphonium (TPP) functionalized mesoporous silica nanoparticles (MSNPs) to target mitochondria (Qu et al., 2015) and fluorescent MSNPs tagged with MT2-AF5p, a peptide targeting membrane-type 2 matrix metalloproteinase on lung cancer cells (Ren et al., 2019). Doxorubicin was the model drug in both examples. Instead of targeting the receptors on the cell membranes, Pan et al. demonstrated using TAT peptide to target nuclei inside cells (Pan et al., 2012). All these examples used mesoporous silica nano-particles because their small size would allow them to internalize into the target cells for drug delivery.

2.4.6 Luminescent Materials

The large pore volume of ordered mesoporous silica has also been used for hosting luminescent materials. There are two different classes of luminescent nanomaterials developed from ordered mesoporous silicas; (1) fluorophores grafting onto mesoporous silicas, and (2) quantum dots formed inside mesoporous silicas.

Fluorophores are commonly used in biological research as markers for imaging and tracking purposes. Some fluorophores are designed for binding proteins and other biomolecules via some simple binding protocols. For example, fluorescein isothiocyanate (FITC) binds to the lysine unit (amine group) of protein molecules following the mechanism shown in Fig. 2.16. This binding chemistry can be exploited to produce a fluorescein-mesoporous silica hybrid. Amine-functionalized mesoporous silica can easily be prepared, via either *in situ* co-condensation or post-synthesis grafting. Therefore direct binding of FITC onto amine-functionalized mesoporous silica is the first to be reported for preparing these fluorescent silica particles (Yang et al., 2005). On the other hand, an FITC-silane can also be prepared by binding FITC with APTES or APTMS and this fluorescent silane can then be grafted onto the pores of nanoporous silicas (Lin et al., 2005). FITC-loaded mesoporous silicas can be used as a probe for monitoring the degradation of polymer scaffolds in tissue engineering (Yang et al., 2005). Another direction is using these fluorescent particles as markers for cells after being internalized (Slowing et al., 2006).

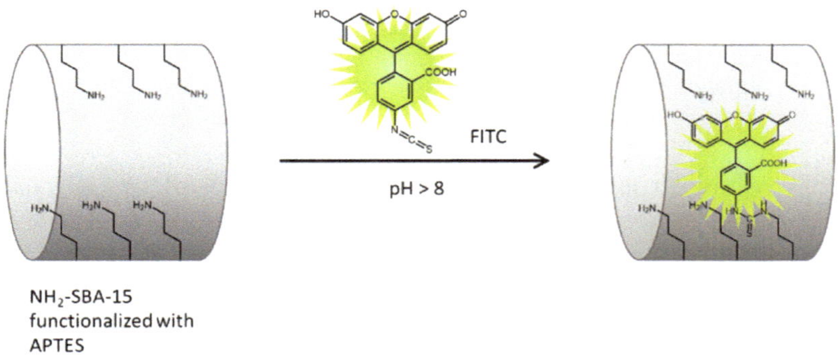

Figure 2.16 Illustration for the mechanism of grafting FITC onto an amine-functionalized SBA-15.

One limitation of fluorophores is that they lose their fluorescence over time due to bleaching. Unlike fluorophores, quantum dots are inorganic fluorescent nanoparticles that are less susceptible to bleaching (see Chapter 1 for details). Quantum dots can also be grown inside ordered mesoporous silicas, as demonstrated by Besson et al. (Besson et al., 2002). In fact, the mesoporous silicas serve two purposes here: (1) as the host and (2) providing a confined space for restricting the growth of quantum dots. This is particularly interesting as the color (emission wavelength) of a quantum dot links to the size of the crystal. Therefore, controlling the growth of these semiconducting crystals is the key to making quantum dots. Figure 2.17 illustrates how CdS quantum dots can be grown inside a mesoporous silica support. Alternatively, pre-made CdS quantum dots can also be hosted inside mesoporous silicas.

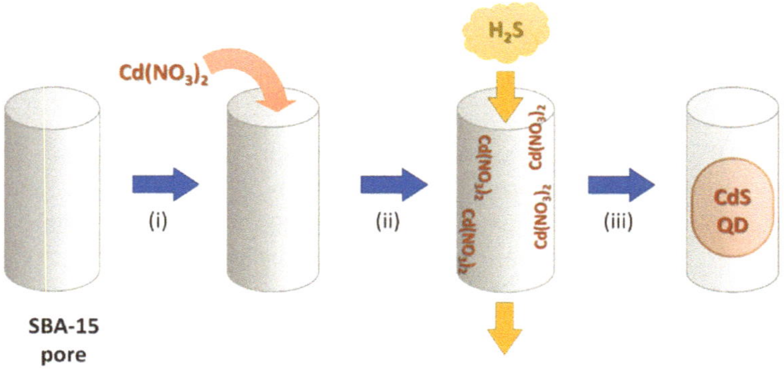

Figure 2.17 Illustration on a synthesis route for growing CdS quantum dots inside a mesopore. Step (i): deposition of $Cd(NO_3)_2$ salt; step (ii) purging H_2S gas through the mesopore; step (iii) growing CdS quantum dots inside mesopore.

Although growing quantum dots inside mesoporous silica is a clever strategy for preparing these luminescent materials, their application is actually limited. One major issue is the quantum dots themselves. Most quantum dots are made from toxic elements, e.g., cadmium, selenium, and thallium, limiting their use in biomedicine. Toxicity is still the major barrier for quantum dots to become a widely used material. On the other hand, there is some recent research making use of this toxicity and exploring the antimicrobial properties of CdS-mesoporous silica composite (Stavitskaya et al., 2022). Such development can take years to mature.

2.5 Non-Siliceous Nanoporous Inorganic Oxides Prepared Using Liquid Crystal Templating

Silica has been the most studied material for liquid crystal templating. This is because of the predominately amorphous structure of silica. Such property is critical because the inorganic lattice has to accommodate the curvature of the liquid crystal template. Crystalline inorganic oxides with a rigid lattice are likely to end up with missing atoms and create vacant sites. These will significantly destabilize the solid overall. Structural collapse is the usual consequence and therefore we do not see often ordered mesoporous materials with crystalline walls.

Nonetheless, mesoporous structures of other inorganic materials can also be found in the literature although they tend to have a lower long-range order (Gu and Schüth, 2014). One of the first examples is the metal mesoporous structure (Attard et al., 1997). However, the preparation procedure for mesoporous metal, e.g., platinum, is very different from that of mesoporous silica. First, a soft liquid crystal template is prepared using surfactant and the metal oxide precursor is then added. The metal nanostructure is formed by a reduction of the formed metal oxide nanostructured materials. Because a true liquid crystal template is required, a much higher surfactant concentration is needed. Also, this method is only limited to small scales.

Mesoporous alumina and titania were also prepared with similar procedures as mesoporous silica. For example, ordered mesoporous alumina prepared with carboxylic acids as surfactant templates were found to have a surface area up to 710 m^2/g (Vaudry et al., 1996). Ordered mesoporous titania was also reported by using alkylamines as templates (Miyake and Kondo, 2001). Since then, many other ordered mesoporous inorganic oxides, mainly transition metal oxides, have been reported using the surfactant templating method. These include Fe_2O_3, Nb_2O_5, ZrO_2, WO_3, and some mixed oxides.

However, many of these mesoporous oxides made from the "soft" liquid crystal templating route tend to have a shorter-range order in their mesostructure compared to their silica counterpart. Therefore, their use on a large scale becomes limited. One route to improving their structure is to use the "hard templating" route for synthesis. This will be discussed in full detail in Chapter 3.

2.6 Concluding Remarks

The discovery of ordered mesoporous silicas has been a turning point for research in nanomaterials in the 1990s. Starting from searching for large-pore catalysts to replace zeolites, these mesoporous silicas have found their use from chromatography to biomedicine. The development also sees changes in direction from metal substitution to inorganic-organic hybrid materials, moving far away from the original concept of "large-pore zeolites." One key feature of these ordered mesoporous silicas is that they allow larger (bio)molecules, or even nanoparticles, to be hosted inside their cavities, leading to the possibility of developing multifunctional materials at the nanoscale. Not many classes of materials can have such property.

The concept of liquid crystal templating was also ground breaking at the time. Few scientists had thought about using the delicate liquid crystals as templates for building porous solids (although very few reported mesoporous materials were actually made from a preformed liquid crystal template, except those developed by Attard et al.). Mesoporous silicas are still by far the most successful examples of liquid crystal templated materials. Success in making other functional oxides with ordered mesoporous structures gave scientists the confidence to explore further in this research direction. More importantly, ordered mesoporous silicas also offer scientists valuable "hard templates" for making other delicate materials. In Chapter 3, we will discuss the use of mesoporous silicas as hard templates for making a wide variety of functional nanomaterials.

References

Armengol, A., Cano, M. L., Corma, A., Garcia, H., and Navarro, M. L. (1995). Mesoporous aluminosilicate MCM-41 as a convenient acid catalyst for Friedel–Crafts alkylation of a bulky aromatic compound with cinnamyl alcohol, *J. Chem. Soc. Chem. Commun.*, **5**, pp. 519–520.

Attard, G. S., Glyde, J. C., and Göltner, C. G. (1995). Liquid-crystalline phases as templates for the synthesis of mesoporous silica, *Nature*, **378**, pp. 366–368.

Attard, G. S., Bartlett, P. N., Coleman, N. R. B., Elliott, J. M., Owen, J. R., and Wang, J. H. (1997). Mesoporous platinum films from lyotropic liquid crystalline phases, *Science*, **278**, pp. 838–840.

Besson, S., Gacoin, T., Ricolleau, C., Jacquiod, C., and Boilot, J.-P. (2002). 3D quantum dot lattice inside mesoporous silica films, *Nano Lett.* **2**, pp. 409–414.

Besson, S., Gacoin, T., Ricolleau, C., Jacquiod, C., and Boilot, J.-P. (2003). Phase diagram for mesoporous CTAB-silica films prepared under dynamic conditions, *J. Mater. Chem.* **13**, pp. 404–409.

Blasco, T., Corma, A., Navarro, M. T., and Pérez Pariente, J. (1995). Synthesis, characterization, and catalytic activity of Ti-MCM-41 structures, *J. Catal.* **156**, pp. 65–74.

Borghard, W. G., Calabro, D. C., Disanzo, F. P., Disko, M. M., Diehl, J. W., Fried, J. C., Markowitz, M. A., Zeinali, M., Melde, B. J., and Riley, A. E. (2009). Characterization and testing of periodic mesoporous organosilicas as potential selective benzene adsorbents, *Langmuir*, **25**, pp. 12661–12669.

Brunel, D., Cauvel, A., Fajula, F., and Direnzo, F. (1995). MCM-41 type silicas as supports for immobilized catalysts, *Stud. Surf. Sci. Catal.* **97**, pp. 173–180.

Burkett, S. L., Sims, S. D., and Mann, S. (1996). Synthesis of hybrid inorganic-organic mesoporous silica by co-condensation of siloxane and organosiloxane precursors, *Chem. Commun.*, **11**, pp. 1367–1368.

Carlsson, N., Gustafsson, H., Thörn, C., Olsson, L., Holmberg, K., and Åkerman, B. (2014). Enzymes immobilized in mesoporous silica: a physical-chemical perspective. *Adv. Colloid Interface Sci.*, **205**, pp. 339–360.

Cheng, C. F., and Klinowski, J. (1996). Synthesis and characterization of the mesoporous galloaluminosilicate molecular sieve MCM-41, *J. Chem. Soc. Faraday Trans.*, **92**, pp. 289–292.

Cheng, C.-F., He, H., Zhou, W., Klinowski, J., Sousa Gonçalves, J. A., and Gladden, L.F. (1996). Synthesis and characterization of the gallosilicate mesoporous molecular sieve MCM-41, *J. Phys. Chem.*, **100**, pp. 390–396.

Chiola, V., Ritsko, J. E., and Vanderpool, C. D. (1971). Process for producing low-bulk density silica, US Patent 3 556 725.

Corma, A., Corell, C., and Pérez-Pariente, J. (1995). Synthesis and characterization of the MCM-22 zeolite, *Zeolites*, **15**, pp. 2–8.

Das, T. K., Chaudari, K., Chandwadkar, A. J., and Sivasanker, S. (1995). Synthesis and catalytic properties of mesoporous tin silicate molecular sieves, *J. Chem. Soc. Chem. Commun.*, pp. 2495–2496.

Díaz, J. F., and Balkus Jr., K. J. (1996). Enzyme immobilization in MCM-41 molecular sieve, *J. Mol. Catal. B: Enzym.* **2**, pp. 115–126.

Doustkhah, E., Lin, J., Rostamnia, S., Len, C., Luque, R., Luo, X., Bando, Y., Wu, K. C.-W., Kim, J., Yamauchi, Y., and Ide, Y. (2019). Development of sulfonic-acid-functionalized mesoporous materials: synthesis and catalytic applications, *Chem – Eur. J.*, **25**, pp. 1614–1635.

Eswaramoorthy, M., Neeraj, and Rao, CNR (1998). High catalytic efficiency of transition metal complexes encapsulated in a cubic mesoporous phase, *Chem. Commun.* pp. 615–616.

Fukushima, Y., and Inagaki, S. (1996). Nano-scale structure control of mesoporous silica, *Mater. Sci. Eng. A*, **217–218**, pp. 116–118.

Gontier, S., and Tuel, A. (1995). Characterization of vanadium-containing mesoporous silicas, *Microporous Mater.*, **5**, pp. 161–171.

Grün, M., Kurganov, A. A., Schacht, S., Schüth, F., and Unger, K. K., (1996). Comparison of an ordered mesoporous aluminosilicate, silica, alumina, titania and zirconia in normal-phase high-performance liquid chromatography, *J. Chromatog. A*, **740**, pp. 1–9.

Gu, D., and Schüth, F. (2014). Synthesis of non-siliceous mesoporous oxides, *Chem. Soc. Rev.*, **43**, pp. 313–344.

Hosein Ghaedi, H., and Ming Zhao, M. (2022). Review on Template Removal Techniques for Synthesis of Mesoporous Silica Materials, *Energy Fuel.*, **36**, pp. 2424–2446.

Han, Y.-J., Stucky, G. D., and Butler, A. (1999). Mesoporous silicate sequestration and release of proteins, *J. Am. Chem. Soc.*, **121**, pp. 9897–9898.

He, Q., Gao, Y., Zhang, L., Zhang, Z., Gao, F., Ji, X., Li, Y., and Shi, J. (2011). A pH-responsive mesoporous silica nanoparticles-based multi-drug delivery system for overcoming multi-drug resistance, *Biomaterials*, **32**, pp. 7711–7720.

He, Y.-Y., Zhang, J.-H., Pu, Q., Xie, S.-M., Li, Y.-X., Luo, L., Chen, X.-X., and Yuan, L.-M. (2019). A novel chiral inorganic mesoporous silica used as a stationary phase in GC, *Chirality*, **31**, pp. 1053–1059.

Huang, S., Hara, K., and Fukuoka, A. (2009). Green catalysis for selective CO oxidation in hydrogen for fuel cell, *Energy Environ. Sci.* **2**, pp. 1060–1068.

Huo, Q., Margolese, D. I., and Stucky, G. D. (1996). Surfactant control of phases in the synthesis of mesoporous silica-based materials, *Chem. Mater.*, **8**, pp. 1147–1160.

Jang, S. H., Kim, M. J., Ko, J. R., and Ahn, W. S. (2005). Catalytic properties of Ti-HMS with high titanium loadings, *Bull. Korean Chem. Soc.*, **26**, pp. 1214–1218.

Junges, U., Jacobs, W., Voigt-Martin, I., Krutzsch, B., and Schüth, F. (1995). MCM-41 as a support for small platinum particles: a catalyst for low-temperature carbon monoxide oxidation, *J. Chem. Soc. Chem. Commun.*, pp. 2283–2284.

Kang, K. K., and Ahn, W. S. (2000). Physiochemical properties of transition metal-grafted MCM-48 prepared using matallocene precursors, *J. Mol. Catal. A: Chem.* **159**, pp. 403–410.

Kang, K. K., Byun, C. S., and Ahn, W. S. (2000). Titanium iso-propoxide grafting on M41S type hosts: catalytic and adsorption study, *Stud. Surf. Sci. Catal.*, **129**, pp. 335–340.

Kim, J. B., and Inui, T. (1996). Synthesis of metal-incorporated mesoporous crystalline silicates for oligomerization of propene, *Catal. Lett.*, **36**, 255–216.

Kloetstra, K. R., and van Bekkum, H. (1995). Base and acid catalysis by the alkali-containing MCM-41 mesoporous molecular sieve, *J. Chem. Soc. Chem. Commun.*, pp. 1005–1006.

Kozhevnikov, I. V., Kloetstra, K. R., Sinnema, A., Zandbergen, H. W., Van Bekkum, H. (1996). Study of catalysts comprising heteropoly acid $H_3PW_{12}O_{40}$ supported on MCM-41 molecular sieve and amorphous silica, *J. Mol. Catal. A: Chem.*, **114**, pp. 287–298.

Kresge, C. T., Leonowicz, M. E., Roth, W. J., Vartuli, J. C., and Beck, J. S. (1992). Ordered mesoporous molecular sieves synthesized by a liquid-crystal template mechanism, *Nature*, **359**, pp. 710–712.

Lin, Y.-S., Tsai, C.-P., Huang, H.-Y., Kuo, C.-T., Hung, Y., Huang, D.-M., Chen, Y.-C., and Mou, C.-Y. (2005). Well-ordered mesoporous silica nanoparticles as cell markers, *Chem. Mater.*, **17**, pp. 4570–4573.

Liu, S., He, H., Luan, Z., and Klinowski, J. (1996). Solid-state NMR studies of the borosilicate mesoporous molecular sieve MCM-41, *J. Chem. Soc. - Faraday Trans.*, **92**, pp. 2011–2015.

Lu, S., Zou, K., Guo, B., Pei, J., Wang, Z., Xiao, W., Zhao, L. (2022). One-step purification and immobilization of thermostable β-glucosidase on Na-Y zeolite based on the linker and its application in the efficient production of baohuoside I from icariin, *Bioorg. Chem.*, **121**, p. 105690.

Macquarrie, D. J. (1996). Direct preparation of organically modified MCM-type materials. Preparation and characterisation of aminopropyl-MCM and 2-cyanoethyl-MCM, *Chem. Commun.*, pp. 1961–1962.

Macquarrie, D. J., Jackson, D. B., Mdoe, J. E. G., and Clark, J. H. (1999). Organomodified hexagonal mesoporous silicates, *New J. Chem.*, **23**, pp. 539–544.

Martínez-Vargas, D. X., Rivera De La Rosa, J., Sandoval-Rangel, L., Guzmán-Mar, J. L., Garza-Navarro, M. A., Lucio-Ortiz, C. J., and De Haro-Del Río, D. A. (2017). 5-Hydroxymethylfurfural catalytic oxidation under mild conditions by Co (II), Fe (III) and Cu (II) Salen complexes supported on SBA-15: synthesis, characterization and activity, *Appl. Catal. A: Gen.*, **547**, pp. 132–145.

Mehnert, C. P., and Ying, J. Y. (1997). Palladium-grafted mesoporous MCM-41 material as heterogeneous catalyst for Heck reactions, *Chem. Commun.*, pp. 2215–2216.

Miyake, Y., and Kondo, T. (2001). Mesoporous titania prepared in the presence of Alkylamine, *J. Chem. Eng. Japan*, **34**, pp. 319–325.

Mohammadnezhad, G., Soltani, R., Abad, S., Dinari, M. (2017). A novel porous nanocomposite of aminated silica MCM-41 and nylon-6: isotherm, kinetic, and thermodynamic studies on adsorption of Cu(II) and Cd(II), *J. Appl. Polym. Sci.*, **134**, p. 45383.

Montalvo, G., Valiente, M., Rodenas, E. (1995). Study of the phase diagram of the CTAB/benzyl alcohol/watersystem, *J. Colloid Interf. Sci.*, **172**, pp. 494–501.

Pan, L., He, Q., Liu, J., Chen, Y., Ma, M., Zhang, L., and Shi, J. (2012). Nuclear-targeted drug delivery of tat peptide-conjugated monodisperse mesoporous silica nanoparticles, *J. Am. Chem. Soc.*, **134**, pp. 5722–5725.

Primo, A., and Garcia, H. (2014). Zeolites as catalysts in oil refining, *Chem. Soc. Rev.*, **43**, pp. 7548–7561.

Qu, Q., Ma, X., and Zhao, Y. (2015). Targeted delivery of doxorubicin to mitochondria using mesoporous silica nanoparticle nanocarriers, *Nanoscale*, **7**, pp. 16677–16686.

Quiton, K. G. N., Lu, M.-C., and Huang, Y.-H. (2021). Synthesis and catalytic utilization of bimetallic systems for wastewater remediation: a review, *Chemosphere*, **262**, p. 128371.

Ray, S., Mapolie, S. F., and Darkwa, J. (2007). Catalytic hydroxylation of phenol using immobilized late transition metal salicylaldimine complexes, *J. Mol. Catal. A: Chem.*, **267**, pp. 143–148.

Rebbin, V., Schmidt, R., and Fröba, M. (2006). Spherical particles of phenylene-bridged periodic mesoporous organosilica for high-performance liquid chromatography, *Angew. Chem. Int. Ed.*, **45**, pp. 5210–5214.

Reddy, K. M., Moudrakovski, I., and Sayari, A. (1994). Synthesis of mesoporous vanadium silicate molecular sieves, *J. Chem. Soc. Chem Commun.*, pp. 1059–1060.

Ren, L., Ma, Z., Li, Q., Zhao, W., Wang, Y., Wang, H., Shen, L., Zhang, C., Fang, X., and Yu, J. (2019). Identifying a membrane-type 2 matrix metalloproteinase-targeting peptide for human lung cancer detection and targeting chemotherapy with functionalized mesoporous silica, *ACS Appl. Bio Mater.*, **2**, pp. 397–405.

Rubin, M. K., and Chu, P. (1990). Composition of synthetic porous crystalline material, its synthesis and use, US Pat. 4,954,325.

Ryu, S. C., Lee, J. H., and Moon, H. (2019). Synthesis of mesoporous silica SBA-15 using a dropwise flow reactor, *Korean J. Chem. Eng.*, **36**, pp. 1410–1416.

Setoguchi, Y. M., Teraoka, Y., Moriguchi, I., Kagawa, S., Tomonaga, N., Yasutake, A., and Izumi, J. (1997). Rapid room temperature synthesis of hexagonal mesoporous silica using inorganic silicate sources and cationic surfactants under highly acidic conditions, *J. Porous Mater.*, **4**, pp. 129–134.

Shinoda, T., Izumi, Y., and Onaka, M. (1995). FSM-16: a recyclable mesoporous acid promoter for meso-tetraarylporphyrin synthesis. *Chem. Comm.*, pp. 1801–1802.

Sigma Aldrich (2022). www.sigmaaldrich.com (accessed 1 Aug 2022).

Slowing, I., Trewyn, B. G., and Lin, V. S.-Y. (2006). Effect of surface functionalization of MCM-41-type mesoporous silica nanoparticles on the endocytosis by human cancer cells, *J. Am. Chem. Soc.*, **128**, pp. 14792–14793.

Stavitskaya, A., Sitmukhanova, E., Sayfutdinova, A., Khusnetdenova, E., Mazurova, K., Naumenko, E., and Fakhrullin, R. (2022). Photoinduced antibacterial activity and cytotoxicity of CdS stabilized on mesoporous aluminosilicates and silicates, *Pharmaceutics*, **14**, p. 1309.

Sutra, P., and Brunel, D. (1996). Preparation of MCM-41 type silica-bound manganese(III) Schiff-base complexes, *Chem. Commun.*, pp. 2485–2486.

Taghizadeh, T., Talebian-Kiakalaieh, A., Jahandar, H., Amin, M., Tarighi, S., and Faramarzi, M. A. (2020), Biodegradation of bisphenol A by the immobilized laccase on some synthesized and modified forms of zeolite Y, *J. Hazardous Mater.*, **386**, p. 121950.

Tatsumi, T., Koyano, K. A., and Igarashi, N. (1998). Remarkable activity enhancement by trimethylsilylation in oxidation of alkenes and alkanes with H_2O_2 catalyzed by titanium-containing mesoporous molecular sieves, *Chem. Commun.*, pp. 325–326.

Tanev, P. T., and Pinnavaia, T. J. (1996). Mesoporous silica molecular sieves prepared by ionic and neutral surfactant, templating: a comparison of physical properties, *Chem. Mater.*, **8**, pp. 2068–2079.

Thomas, J. M., Raja, R., Johnson, B. F. G., Hermans, S., Jones, M. D., and Khimyak, T. (2003). Bimetallic catalysts and their relevance to the hydrogen economy, *Ind. Eng. Chem. Res.*, **42**, pp. 1563–1570.

Tuel, A., and Gontier, S. (1996). Synthesis and characterization of trivalent metal containing mesoporous silicas obtained by a neutral templating route, *Chem. Mater.*, **8**, pp. 114–122.

Vaudry, F., Khodabandeh, S., and Davis, M. E. (1996). Synthesis of pure alumina mesoporous materials, *Chem. Mater.*, **8**, pp. 1451–1464.

Vallet-Regi, M., Rámila, A., Del Real, R. P., and Pérez-Pariente, J. (2001). A new property of MCM-41: drug delivery system, *Chem. Mater.*, **13**, pp. 308–311.

Van Rhijn, W. M., De Vos, D. E., Sels, B. F., Bossaert, W. D., and Jacobs, P. A. (1998a). Sulfonic acid functionalised ordered mesoporous materials as catalysts for condensation and esterification reactions, *Chem. Commun.*, pp. 317–318.

Van Rhijn, W., De Vos, D., Bossaert, W., Bullen, J., Wouters, B., Grobet, P., and Jacobs, P. (1998b). Sulfonic acid bearing mesoporous materials as catalysts in furan and polyol derivatization, *Stud. Surf. Sci. Catal.* **117**, pp. 183–190.

Vaudry, F., Khodabandeh, S., and Davis, M. E. (1996). Synthesis of pure alumina mesoporous materials, *Chem. Mater.*, **8**, pp. 1451–1464.

Wu, S., Li, F., Xu, R., Wei, S., and Li, G. (2010). Synthesis of thiol-functionalized MCM-41 mesoporous silicas and its application in Cu(II), Pb(II), Ag(I), and Cr(III) removal, *J. Nanoparticle Res.*, **12**, pp. 2111–2124.

Yanagisawa, T., Shimizu, T., Kuroda, K., and Kato, C. (1990). The preparation of alkyltrimethylammonium-kanemite complexes and their conversion to microporous materials, *Bull. Chem. Soc. Japan*, **63**, pp. 988–992.

Yang, Y., Yiu, H. H. P., and El Haj, A. J. (2005). On-line fluorescent monitoring of the degradation of polymeric scaffolds for tissue engineering, *Analyst*, **130**, pp. 1502–1506.

Yiu, H. H. P. Wright, P. A., and Botting, N. P. (2001a). Enzyme immobilisation using SBA-15 mesoporous molecular sieves with functionalised surfaces, *J. Mol. Catal. B Enzym.*, **15**, pp. 81–92.

Yiu, H. H. P., Botting, C. H., Botting, N. P., and Wright, P. A. (2001b). Size selective protein adsorption on thiol-functionalised SBA-15 mesoporous molecular sieve, *Phys. Chem. Chem. Phys.*, **3**, pp. 2983–2985.

Yu, C., Yu, Y., Miao, L., and Zhao, D., (2001). Highly ordered mesoporous silica structures templated by poly(butylene oxide) segment di- and tri-block copolymers, *Microporous Mesoporous Mater.*, **44–45**, pp. 65–72.

Zhang, L., and Ying, J. Y. (1997). Synthesis and characterization of mesoporous niobium-doped silica molecular sieves, *AIChE J.*, **43**, pp. 2793–2801.

Zhao, D., Huo, Q., Feng, J., Chmelka, B. F., and Stucky, G. D. (1998a). Nonionic triblock and star diblock copolymer and oligomeric sufactant syntheses of highly ordered, hydrothermally stable, mesoporous silica structures, *J. Am. Chem. Soc.*, **120**, pp. 6024–6036.

Zhao, D., Feng, J., Huo, Q., Melosh, N., Fredrickson, G. H., Chmelka, B. F., and Stucky, G. D. (1998b). Triblock copolymer syntheses of mesoporous silica with periodic 50 to 300 angstrom pores, *Science*, 279, pp. 548–552.

Zhao, J., Gao, F., Fu, Y., Jin, W., Yang, P., and Zhao, D. (2002). Biomolecule separation using large pore mesoporous SBA-15 as a substrate in high performance liquid chromatography, *Chem. Commun.*, **2**, pp. 752–753.

Zhu, G., Jiang, D., Yang, Q., Yang, J., and Li, C. (2007). Trans-(1R,2R)-Diaminocyclohexane-functionalized mesoporous organosilica spheres as chiral stationary phase, *J. Chromatog. A*, **1149**, pp. 219–227.

Chapter 3

Extended Materials from Ordered Mesoporous Silica: Mesocellular Foams and Non-Siliceous Mesoporous Materials

Humphrey H. P. Yiu and Gregor Sneddon

School of Engineering and Physical Sciences, Heriot-Watt University, Edinburgh, EH14 4AS, UK

h.h.yiu@hw.ac.uk

3.1 Introduction

In Chapter 2, we have seen the usefulness of the long-range order of templated mesoporous silicas and their versatility. However, there are three limitations of mesoporous silica. Its pore size is limited to several nm wide, which could still be too narrow for many macromolecules, including many biomolecules. Its hydrophilic nature may not be always suitable for guest molecules. Its lack of chemical activities means that additional functionalization is necessary for many desired applications. Therefore, scientists worked on the possibility to exploit the success of ordered mesoporous silicas to develop (1) materials

Nanotemplating: Engineering Nanocomposite Materials for the Future
Humphrey H. P. Yiu

ISBN 978-981-4316-92-7 (Hardcover), 978-1-003-50999-8 (eBook)
www.jennystanford.com

that can accommodate larger molecules by enlarging the pore size, (2) hydrophobic materials that still process an ordered mesoporous structure, and (3) materials that are intrinsically active without further modifications. In this chapter, we will go further to introduce three families of materials that are related to ordered mesoporous silicas.

3.2 Mesocellular Foams

One limitation of templated mesoporous silica is that their pore size is limited to the dimension of the liquid crystal templates, typically 2–5 nm wide. To expand this further, initially, scientists thought we could simply use larger surfactant molecules and the pore size would become larger. Unfortunately, this is only true to a certain extent. For example, SBA-15 has a larger pore size than MCM-41 by about 2-fold but the surfactant Pluronic P123 used for SBA-15 synthesis is already at the molecular weight of 5700 g/mol (or Da). If the size of the template molecules increases further, e.g., Pluronic F127 (12,500 g/mol), the pore size of the mesoporous materials formed (SBA-16) is, in fact, smaller but the pore structure is completely different. Therefore, increasing the size surfactant template is not going to help to increase the pore size to accommodate much larger molecules.

Since these micelles are essentially an oil-in-water emulsion, and the surfactant molecules are capable of locking the oil phase in micelles, increasing the volume of the oil phase should in theory increase the size of the micelles, as shown in Fig. 3.1. Hence, scientists started to investigate the viability of using oily "expander" to increase the pore size of mesoporous silica.

Indeed, this theory was proven to be workable in one of the first mesoporous silica papers (Beck et al., 1992). In Beck's work, an MCM-41 sample was shown to have a larger pore by adding 1,3,5-trimethylbenzene (or mesitylene) as "oil" during the synthesis. However, there is only one such sample reported in the paper, and not a lot of follow-up research for the next few years. This is partly because the oil-to-surfactant ratio seemed to be quite delicate and the pore size was not enlarged much (up to 10 nm as it was claimed). More importantly, the materials seem to start losing their long-range order in pore structure, which was a definite negative aspect at the time, showing no XRD peaks.

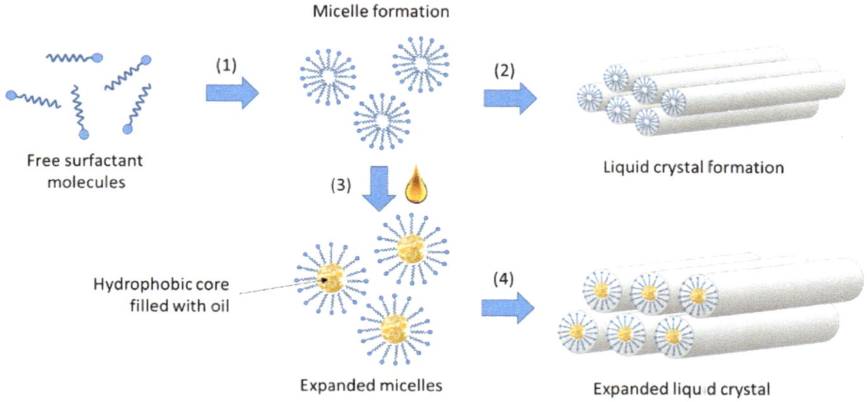

Figure 3.1 Illustration showing pore expanding on mesoporous silica by the addition of an oil. (1) formation of micelles when the surfactant concentration exceeds CMC1 (critical micelle concentration 1), (2) formation of the liquid crystal when the surfactant concentration exceeds CMC2, (3) addition of oil to expand the micelle size, (4) formation of a liquid crystal of expanded micelles.

The situation changed when Stucky's research group expanded their work by adding mesitylene in an SBA-15 synthesis. (Schmidt-Winkel et al., 1999). The "disordered" materials were further studied in detail and they were found to have a "bubble-like" 3-D structure, with windows and cages interconnecting to each other (see Fig. 3.2). More important, the cage and window sizes were linked to the amount of mesitylene added. The oil-to-surfactant ratio can also go as far as 1:1 by mass. Due to the major differences in pore structure to SBA-15 and indeed that they have no long-range order at all, these materials were given a different name, Mesocellular Foams, more commonly known as MCF.

The key feature of MCF materials is their pore size is now as large as 10 nm, wide enough to accommodate many biological molecules including most enzymes, antibodies and some polymer molecules (Han et al., 1999). In Han's work, large proteins were used to prove that the materials can allow macromolecules to enter. As a result, the development of MCFs tends to be directed towards biological applications.

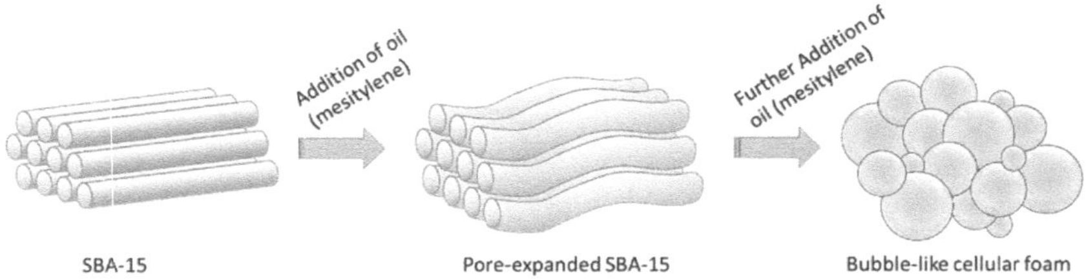

Figure 3.2 Illustration depicting the transition from SBA-15 to MCF by the addition of an oil (mesitylene in most cases).

3.2.1 Synthesis and Characterizations of MCFs

Although the pore expansion using oil can be applied to all micelle-templated mesoporous silicas, MCFs, which are formed by expanding the pores of SBA-15, are the only widely studied materials developed from ordered mesoporous silicas. Except for an additional oil component, the method for synthesizing MCFs is essentially the same as that of SBA-15, even including the reaction conditions and chemicals. To start with, synthesis for SBA-15 requires the solvation of the surfactant template at a specific concentration in an acidified medium (see Chapter 2). The same applies to MCF synthesis. However, once the surfactant solution is made, the oil expander, e.g., mesitylene, is added to make a light emulsion. This could take time for all the oil to be "solvated" inside the micelles; the higher the oil content, the longer it takes. For example, for an MCF with a surfactant-to-oil ratio of 1:1 by weight, it will take roughly a day for complete solvation. This can be visually checked on any "shiny" oil layer at the surface of the solution. Once all oil is solvated, the silica precursor, usually TEOS, can then be added, the same as in SBA-15. This is followed by the hydrothermal and calcination steps, again the same as SBA-15 synthesis.

For SBA-15, functionalization with organic groups can be performed via either *in situ* or post-synthesis routes (see Chapter 2). However, for MCFs, *in situ* is rarely performed because the addition of another silica source (e.g., APTES) will only make the synthesis too complicated and the delicate balance of foam formation may be difficult to retain. Hence, post-synthesis functionalization is generally preferred. The functionalization step using organosilanes is again essentially the same as other mesoporous silicas. Indeed, once the porous materials are formed, no matter if it is SBA-15 or MCF, the surface chemistry is basically the same because they are both amorphous silicas.

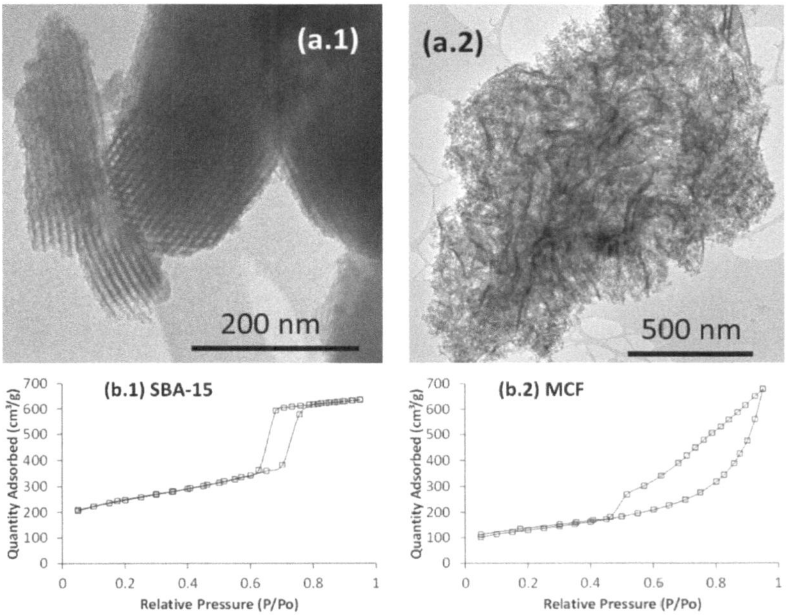

Figure 3.3 Structural comparison of SBA-15 and MCF with (a) TEM images and (b) N_2 adsorption isotherms.

In general, the same characterization techniques for SBA-15 are to be used for MCFs as well, except that XRD is unlikely to be applied due to the lack of long-range ordered structure. Commonly, only TEM and N_2 adsorption data are presented for MCF materials. TEM can confirm the foam structure of MCFs while N_2 adsorption gives essential structural parameters such as pore size, specific pore volume and BET-specific surface area. Generally, the pore size of MCFs ranges from 5–10 nm and the specific pore volume can be as high as 2 cm^3/g. These are the results of adding an oil expander. However, the BET-specific surface area may not increase much from their SBA-15 cousin, usually at around 500 m^2/g. Figure 3.3 shows some typical differences between the TEM and N_2 adsorption isotherm of SBA-15 and an MCF material.

3.2.2 Applications of MCFs

The main reason for the development of an "expanded version" of mesoporous silica is to accommodate macromolecules that are simply too large for mesoporous silica to cope with. Therefore, MCF silicas find their uses in biology as biomolecules are usually large (> 5 nm).

3.2.2.1 Separation of biomolecules

As developed initially, MCF materials were found to be able to "sort" biomolecules based on either molecular sizes. Han et al. demonstrated that MCF can allow larger protein (conalbumin, mw = 77 kDa) to be bound onto its large pore but not observed with SBA-15 (Han et al., 1999). It is proven evidence of MCF's superiority over SBA-15 for systems involving large molecules. Indeed, this formed the blueprint for most work in this research area.

Further development in this area lies in the practicality of MCF in analytical instrumentation setups. For example, Han, Lee and Ying prepared MCF spheres and tested their sorting ability on polystyrene standards using an HPLC system, as a demonstration of size exclusion chromatography (SEC) (Han et al., 2007). The results showed a remarkable correlation between the pore size of MCF and the molecular weight of the polystyrene analyte. The use of MCF can reduce the column length and analysis time. A similar study was carried out using amine-functionalized MCF particles for the analysis of protein using an HPCL system (Sun et al., 2016). The system packed with amine-functionalized MCF gave a higher resolution than a commercial amine column, which could not separate horseradish peroxidase (44 kDa) from lysozyme (14–15 kDa). The same system was also used to separate anilines and polycyclic aromatic hydrocarbons (PAH) and similar results were observed.

One problem in this area of application is that although MCFs have an expanded porous structure from SBA-15, they lack

SBA-15's uniform pore size distribution. Therefore, they are less specific in sorting biomolecules. Nonetheless, it is still better than other non-templated options such as porous cellulose, which is widely used in dialysis and filtration for biomolecules.

3.2.2.2 Enzyme immobilization

In Chapter 2, we have introduced the use of mesoporous silica (mostly SBA-15) for enzyme immobilization. However, as we also discussed, SBA-15 can only allow molecules smaller than 5 nm in diameter (ca. 40 kDa) to enter their pores, and many enzymes are larger than this range. Moreover, even if those enzymes can enter the pores, the blockage is likely to occur, leading to the diffusion of substrates and products being severely hampered. This is very important to catalysis (enzymatic reactions are simply biocatalysis) as it slows down the whole process greatly. Such reduction in catalytic activities is considered to be negative to many industries. Finally, the pore structure of SBA-15 is a 2-D hexagonal pack. When blockage does occur, there is no alternative route for the diffusion of substrate and product molecules. The use of other 3-D porous structures, including SBA-16 and FDU-1, can circumvent this problem but their narrow window size makes fewer choices of enzymes to be accommodated. MCFs become the undisputed winner for this application.

There are many examples of using MCFs as a carrier for enzymes, but the principles are generally the same. For example, glucose oxidase (GOx) has been immobilized onto MCF and compared with using SBA-15 as the carrier (Jung and Hartmann, 2010). An oxidation reaction of indole to oxindole was used to test these GOx systems. With enlarged pores, leaching of GOx from MCF saw the enzymatic activity diminished in 12 h. However, when GOx was crosslinking inside the pores with CLEA (cross-linked enzyme aggregate), the activity was shown up to a 10-fold increase in TOF (per mass of enzyme) when compared with GOx inside SBA-15. In a different study, MCF was used to immobilize laccase for the removal of tetracycline (Zdarta, et al., 2020). Laccase immobilized on MCF has shown improved activity over free laccase and is still active (~80%) after 30 days, which has seen the activity of free enzyme decline to only 40% of its original activity. Also, the inclusion of copper chloride in the MCF support has seen the activity improving further as ~90% of the activity was retained after 10 cycles. Other examples include β-glucosidase (Zlateski et al., 2015), α-amylase, (Agustian and Hermida, 2019), and cellulase (Kannan and Jasra, 2011).

3.2.2.3 Support for functional polymers

Functional polymers are essential in materials science as they are versatile for various applications. However, most functional polymers suffer from a low surface area-to-volume ratio. In a way, many "functionalities" on the polymer are being shielded inside. This is particularly difficult for molecules/analytes to reach if the polymer is used as a solid (e.g., as the stationary phase). This has been demonstrated using SBA-15 when the mesoporous support was used to host polyaniline, a conductive polymer. However, the difference between this application and hosting an enzyme is that the hosted conductive polymer would have little interaction with other molecules (e.g., a substrate). Therefore, problems due to blockages in pores post immobilization became unimportant. However, if access for "foreign molecules" to this functional polymer is essential to a specific application, SBA-15 becomes unsuitable.

One example of MCFs hosting a functional polymer has been shown when they were used to support chitosan for CO_2 adsorption (or carbon capture). Chitosan is in fact a natural polymer with a polysaccharide structure and amine groups on its backbone. It is derived from shells and is one major polymer waste worldwide. Finding a useful application for this waste is an attractive prospect.

Regenerable absorbents of CO_2 are highly sought after in recent years due to growing concern about the ever-increasing CO_2 level in the atmosphere. Carbon capture technology is considered to be a solution. However, the current carbon capture technologies are still highly energy intensive, mostly due to the regeneration of adsorbents (e.g., CaO) or absorbents (sodium hydroxide and amine solutions) (Sneddon et al., 2014). As such, it is difficult to achieve genuine negative carbon emission, or even *Net Zero*. Generally, an adsorbent need to be regenerated at a temperature below 80°C to be considered a "workable" adsorbent.

There is a common feature in most CO_2 adsorbents/absorbents; they are all of a basic (or alkaline) nature, e.g., sodium hydroxide or calcium oxide, to react chemically with CO_2 (a weakly acidic gas). This is because there is only 8–15% CO_2 in flue gas from power stations, while this percentage is even lower in many other sources. Other major components (N_2, O_2 and H_2O) are all neutral and therefore a basic adsorbent/absorbent is the most effective. However, strong bases require high energy to regenerate due to the strong binding between chemically adsorbed/absorbed CO_2. For example, CaO solid needs to be heated at 700°C for regeneration ($CaCO_3 \rightarrow CaO + CO_2$), which is likely to consume much more energy and the whole process may still have an overall positive carbon emission. Using an aqueous solution of amine can reduce this regeneration temperature to below 150°C but the majority of the energy (estimated to be 70–90%) was consumed to evaporate water, which is about 80–90% of the solution. Evaporation of water is a very energetic process but water is not responsible directly for CO_2 interaction. Hence the process is also deemed to be energy inefficient.

Since chitosan has intrinsically basic amine groups in the structure so no further functionalization or modification is required for CO_2 adsorption. Chitosan had already been considered for CO_2 adsorption many years ago by a Japanese group (Yoshida et al., 2002), but the low specific surface area is not favorable. Supporting chitosan on a porous amberlite resin had shown improvement in CO_2 adsorption capacity but not enough to be considered a viable solution. In another research work, chitosan was supported on MCF materials made with different mesitylene:surfactant ratios (0.3:1–1.0:1). The results were compared with that supported on SBA-15 and showed significant improvement from that on a resin. (Sneddon et al., 2015).

In the same area of application, MCF supports were also used to host aminated PVC for carbon capture (Sneddon et al., 2017). Again, this has demonstrated the use of the MCF porous structure. More significantly, it demonstrated that polymer wastes can be used for tackling another environmental problem. Other examples include supporting polyethyleneimine (PEI), tetraethylenepentamine (TEPA) (Feng et al., 2013,) and CaO crystallites (Gao et al., 2019) for CO_2 capture. However, TEPA is not a large molecule and is accessible to mesopores of SBA-15. Hence, the advantage of using MCF's large-pore structure does not stand out here. Regarding supported CaO, the solid CaO crystallites are likely to block the mesopores of SBA-15 and the use of MCF support could avoid this problem. The real issue lies in the base CaO itself due to the high energy requirement upon regeneration, as mentioned previously. The use of MCF supports does not solve the problem associated with a high energy requirement for adsorbent regeneration.

3.2.2.4 Biomedical applications

Similar to SBA-15 and other silica-based nanoporous materials, MCFs are considered to be biocompatible and non-cytotoxic. Therefore, they have been tested for use in biomedical applications. One key biomedical area would consider these high-surface-area solid-state materials to be used as carriers for drug delivery. For example, simvastatin was loaded onto an MCF carrier and the release profile was studied in a simulated intestinal fluid (Zhang et al., 2011). Simvastatin was chosen because of its poor solubility in water. When compared with SBA-15, MCF showed a higher loading efficiency (at 37.5%) and a faster release rate. This is likely to be related to the pore size of carriers. This research group also studied using MCF as a host for telmisartan, and similar results were found (Zhang et al., 2010). Indeed, a later study used a drug release model to study the release behavior of MCF carriers (Liu et al., 2019). Release in the early stage (1 h), or initial rate of release, is solely affected by the drug loading while cumulative release long term (24 h) is influenced by both drug loading and pore size.

As mentioned in the previous section, functional polymers have been incorporated into the MCF host to increase the surface area of these guest polymers. Similar polymer-MCF composites have also been tested for drug delivery. For instance, MCF was used to host paliperidone as this composite was further encapsulated into poly(lactic acid) PLA and poly(lactide-co-glycolide) PLGA. An outmost coating of thiolated chitosan was used to control intestinal bursts (Nanaki et al., 2017).

3.3 Mesoporous Carbons: CMK-Type Materials

The complex, delicate structure of nanoporous materials has not been unnoticed for use as a template. A second generation of nanoporous materials can indeed be templated from a nanoporous solid (e.g., mesoporous silica). However, in the beginning, scientists were struggling with a couple of technical problems; how to remove the silica template without damaging the newly formed materials; and, how can the newly formed materials retain their structure without collapsing after template removal. These issues seem trivial now but, they have been the major obstacles in the early development of such new materials.

3.3.1 Development of Mesoporous Carbons

The earliest example of using ordered mesoporous silicas as templates for preparing nanoporous materials was ordered mesoporous carbon developed by Ryoo's group in Korea (Ryoo et al., 2001). This carbon material is formed by anaerobic pyrolysis of sucrose inside an ordered mesoporous silica template, which was then removed using HF or a strong base such as NaOH once the carbon materials were formed. Three mesoporous silica templates (MCM-41, MCM-48 and SBA-15) were used in this first attempt to make ordered mesoporous carbon. Interestingly, they have shown very different results. First, when MCM-41 was used as the template, once the silica template was removed, the structure of this newly formed carbon material collapsed due to the lack of interconnection between the pores in the MCM-41 framework.

When MCM-48, which has an *Ia3d* 3-D porous structure, was used as a template, the problem of structural collapse was solved. Due to the interconnected pore system of MCM-48, the formed carbon framework, named CMK-1, retained its structure upon template removal (Joo et al., 2001). Powder XRD analysis on CMK-1 also revealed its structure being the inverse replicate of MCM-48. This is a very important breakthrough in nanotemplating as it is a true example of a second generation of ordered mesoporous materials.

One disadvantage of MCM-48 templating is the difficulty in tailoring the ordered pores size. Hence there is a very limited variation for CMK-1. This work then moved forward to an SBA-15 replicate in order to prepare a nanoporous carbon structure of a larger unit cell. This SBA-15 template is unique due to its connecting micropores and small mesopores in the walls of large-pore channels of SBA-15. To many scientists' surprise, unlike using MCM-41 as the template, this inverse SBA-15 replicate, called CMK-3,

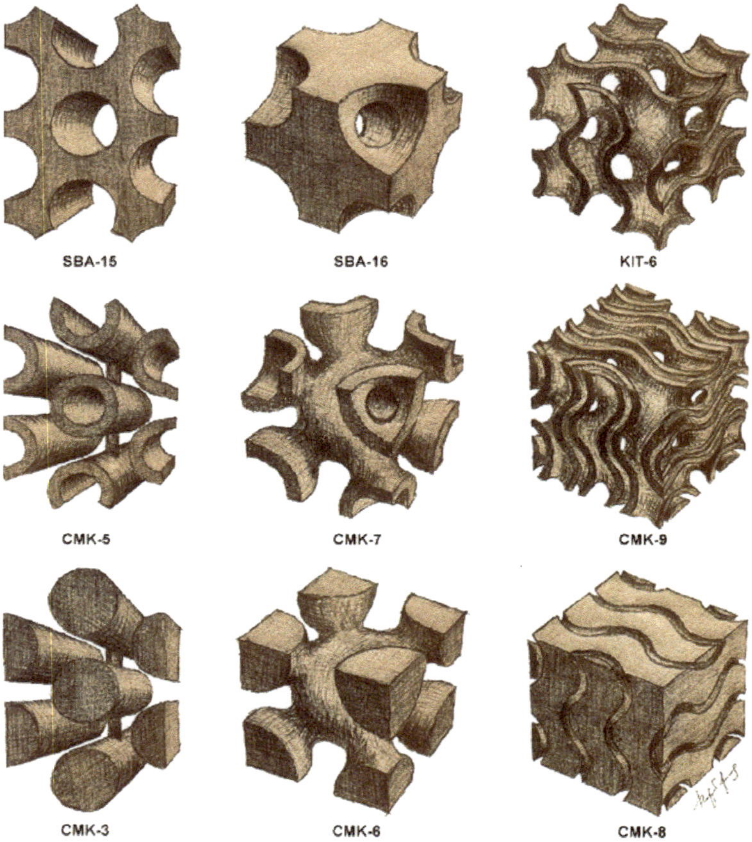

Figure 3.4 Artistic illustration of the structure of several CMK materials (Janus et al., 2022).

retained its structure upon template removal. This is because the carbon formed inside the secondary nanoporous structure of SBA-15, or the disordered intra-pore connective channels, pillars the whole carbon network. There are two significances here; (1) CMK-3 is the first 2-D *p6* nanoporous carbon structure; and (2) CMK-3 is a proof for the inter-pore connective channels in the SBA-15 structure. Since then, there have been other attempts for the synthesis of several templated nanoporous carbon materials. If you wonder what these CMK-x notations are, there is a comprehensive article listing all these CMK materials in full with excellent drawings as shown in Fig. 3.4 (Janus et al., 2022).

3.3.2 Synthesis of CMK-Type Mesoporous Carbons

As mentioned in the previous section, the first successfully prepared ordered mesoporous carbon (CMK-1) was synthesized using MCM-48 silica as a template. The porous system of MCM-48 was then filled with sucrose in the presence of sulfuric acid and the mixture was dried at 100°C, subsequently at 170°C followed by repeating the earlier process once. This step is aimed to dehydrate the sucrose with sulfuric acid as the drying agent. The sucrose filling was eventually carbonized at 900°C under an inert atmosphere, usually N_2. Once the sample was carbonized, the MCM-48 silica template can then be removed by dissolution in a solution of ethanol and NaOH at 100°C. Hydrogen fluoride solution (HF) can also be used for the dissolution but it is a much more dangerous chemical and now it is a less preferred option due to safety reasons. Usually, up to 90% of the silica template can be removed with this method. Once dried in the air, this CMK-1 sample is now ready for use. There are variations in the temperature of carbonization and the conditions for silica template removal. However, the principles are essentially the same.

The synthesis method for CMK-3 from an SBA-15 silica template is basically the same as that of CMK-1 except the template being used is different. We can see that there are various methods for the synthesis of mesoporous carbons with desired pore size distributions via the templating procedures. The techniques developed for synthesizing ordered mesoporous carbons are straightforward and cost-efficient. For example, the carbon source (e.g., sucrose) and the dissolution alkali (NaOH/ethanol) are cheap and widely available chemicals. CMK-3 carbon, one of the highly ordered mesoporous carbons, is particularly promising due to the inexpensive and easily synthesized SBA-15 template and its easily tailored pore wall thickness that allow pore size tailoring. Recent research on the synthesis of MCM-48 (Zhang et al., 2012) had allowed the improvement in the cost-effective and custom-tailored syntheses of these templates. In general, the templates must exhibit a 3-D pore structure for the synthesis of ordered mesoporous carbon in order to prevent structural collapse upon template removal.

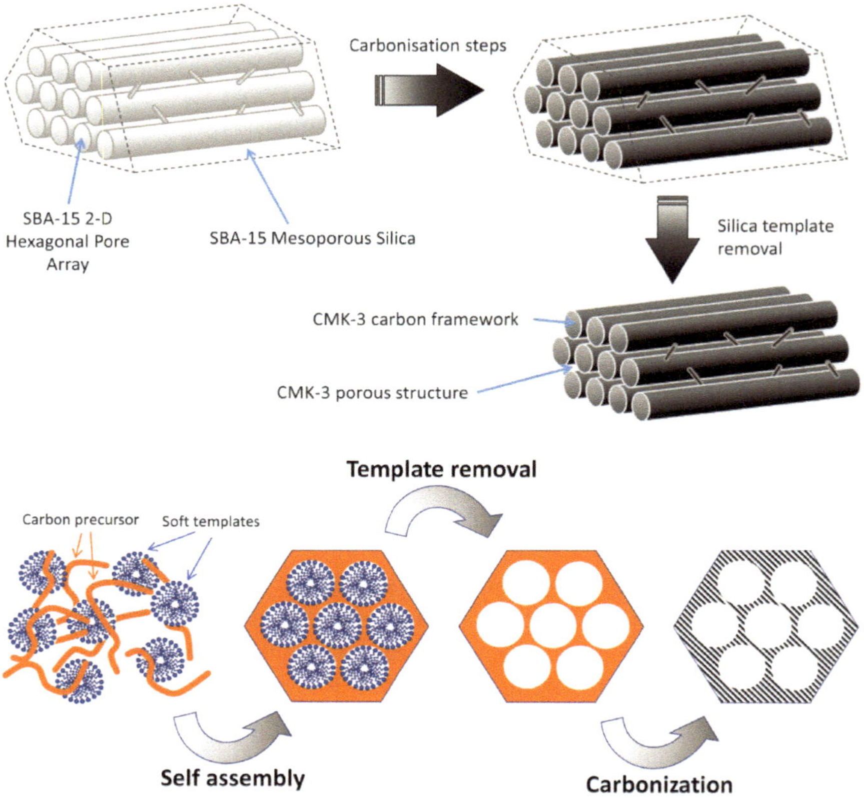

Figure 3.5 Synthesis methods for mesoporous carbon via (a) hard-template and (b) soft-template pathways.

This synthesis for mesoporous carbon via silica-based "hard" template, however, has a clear disadvantage as the template removal using alkaline treatment is still a time-consuming process. Also, unwanted residual silica could be left in the pores of the resultant carbon materials. Such contaminants can reduce the conductivity of the carbon materials or block the porous structure. Hence, another alternative method has been invented that is based on a soft template via a self-assembling block copolymer surfactant and a carbon precursor. The advantage presented here is that the harsh template removal regime would no longer be required and a potentially better reproductivity can be achieved via the use of an aqueous route. In contrast, ordered mesoporous carbons based on hard templates are easier to be graphitized, which can be difficult for soft-templated-based ones. Meanwhile, the soft-templating route offers stability in harsh thermal and oxidation treatments for functionalization (Xin et al., 2015). Figure 3.5 demonstrates a summary of synthesis via hard template (a) and soft template (b) (Eftekhari and Fan, 2017). As an extension of mesoporous silica, hard-templated mesoporous carbon is still our focus in this chapter.

3.3.3 Characterization of Mesoporous Carbons

In terms of meso-scaled structure, mesoporous carbons can be similar to mesoporous silicas, because they are essentially replicates. If the materials were prepared aimed at adsorption/environmental applications, the primary parameter needed to find out is specific surface area and porosity. Naturally, an adsorbent with a high surface area will have a high adsorption capacity for an adsorbate. Specific surface area, typically BET for a mesoporous solid, can be measured using the nitrogen adsorption method, which has been introduced in detail in Chapter 2. The experiment setup and the principle are essentially the same as that for mesoporous silicas. Figure 3.6d shows a typical N_2 adsorption isotherm for a CMK-3 material. TEM also offers examinations on the structure of these carbonaceous materials at the nanoscale. Figure 3.6c shows the typical structure of a CMK-3 sample, which is notably different from an activated carbon (Fig. 3.6a) and carbon nanofiber (Fig. 3.6b).

Regarding surface chemistry, the major difference between mesoporous carbons and mesoporous silicas is the hydrophobicity/hydrophilicity. Carbonaceous materials are generally hydrophobic while silicas are known to be hydrophilic. Therefore, as an adsorbent, mesoporous carbons are preferred for hydrophobic adsorbates such as methane, benzene and other hydrocarbons. To measure the hydrophobicity of a material, "contact angle" measurement is usually employed. Generally, a small droplet of water is dropped onto a clean surface of the sample and the "contact angle" (see Fig. 3.7) is measured by taking a picture of the droplet. The larger the contact angle, the more hydrophobic the surface is.

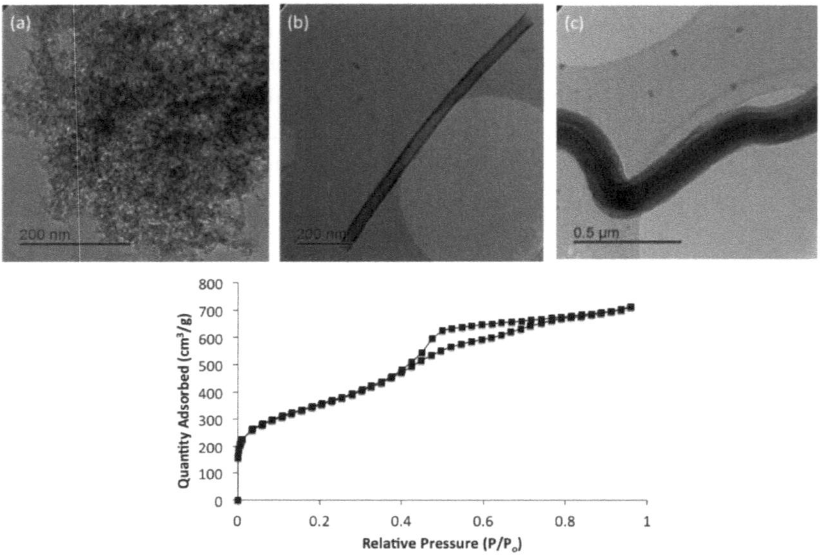

Figure 3.6 TEM images of porous carbon materials (a) activated carbon, (b) carbon nanofiber, and (c) CMK-3 mesoporous carbon. (d) Adsorption isotherm of CMK-3, showing the hysteresis loop.

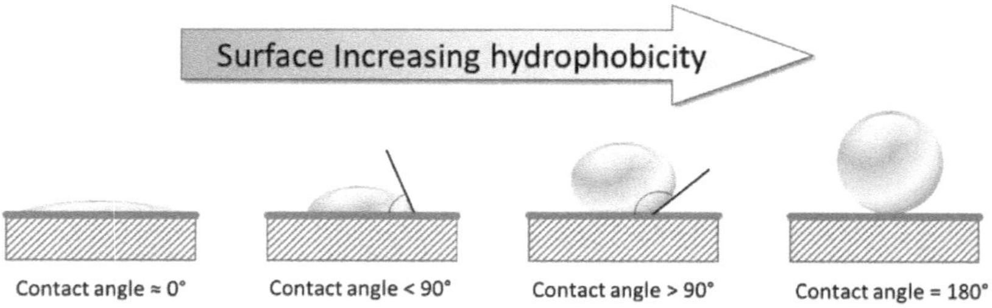

Figure 3.7 Contact angle measurement and its relation to the hydrophobicity of a surface.

Unlike mesoporous silicas, in most research studies, mesoporous carbons were used as they are without further surface functionalization. Hence, there was no need to use vibrational spectroscopy (FTIR) for examining the surface functional groups of mesoporous carbon. In some cases, mesoporous carbons were oxidized using concentrated nitric acid (conc. HNO_3) in order to generate carboxylic acid groups and other groups (including phenolic) on the surface. This is a common method to generate sites for attaching functional molecules or to reduce the hydrophobicity of the mesoporous carbons. FTIR could be used for identifying these groups.

More specifically, Raman spectroscopy is widely used in the characterization of carbon nanomaterials, in particular graphene and carbon nanotubes, for studying the carbon–carbon bondings within the structure. In general, the carbon atoms in both "pure" graphene and carbon nanotube structure are in sp^2 hybridization, with one carbon atom bound with three other carbon atoms, forming the famous hexagonal pattern. Based on the sp^2 signal in a Raman spectroscopy, information about the crystallite size of the sp^2 phase can be provided (Jorio et al., 2011). However, more importantly, any sp^3 hybridization, considered to be structural defects, can be detected. Moreover, any chemical impurities in the carbon structure will also appear in the Raman spectrum. To our focus here, mesoporous carbon is far from a pure sp^2 structure, with a significant amount of amorphous carbon. Raman spectroscopy can help to determine the quality of the mesoporous carbon samples.

Another key property of mesoporous carbons to study is their electrochemical property. Since many mesoporous carbons were prepared for their electrical applications as capacitors or for the construction of fuel cell electrodes, a full understanding of the electrochemical properties is crucial. Measurements such as electrical conductivity and capacitance, as well as cycling durability, are also commonly carried out for mesoporous carbons and their derivative materials (Yan et al., 2020).

3.3.4 Applications of Ordered Mesoporous Carbons

There are numerous potential applications where carbonaceous surface materials would be attractive such as the adsorption of large hydrophobic molecules, chromatographic separations, catalysis, electrochemical double-layer capacitors, and lithium batteries. In these cases, it would be advantageous to have wider pores preferably in the mesopore range. Therefore, the synthesis of mesoporous carbons has recently been of considerable interest. In many areas of modern science and technology, porous carbonaceous materials are important in the application of water and air purification, gas separation, catalysis, chromatography, and energy storage.

3.3.4.1 Environmental applications of ordered mesoporous carbons

High surface area, porous carbonaceous materials such as activated carbon and carbon nanotubes are considered to be good adsorbents for many organic pollutants due to their hydrophobic nature. Carbon nanotubes are still too expensive to be practical as commercial adsorbents. The cheaper activated carbon materials are still preferred adsorbents for water purification. Indeed, they are even widely used in domestic water purifiers. That does not stop scientists from searching for new carbonaceous materials for this application. Ordered mesoporous carbons were therefore tested.

It is common to use an organic dye as a model pollutant for testing the potential of an adsorbent for water purification. The main reason is that the color of a dye is easily detectable using a simple colorimetric method such as UV-vis spectrometry. Also, the organic nature of these dyes can resemble the surface chemistry of an organic pollutant. For example, the performance of the ordered mesoporous carbon (CMK-3 type) as an adsorbent in the removal of dyes from aqueous solutions has been studied with commercial activated carbon used for comparison (Böke et al., 2013). In this study, the adsorption capacity of carbon materials was tested using three different dyes: acid orange 8, methylene blue, and rhodamine B solutions at various pH. The results suggest that pH plays a strong role in the adsorption capacity of mesoporous carbon for these dyes due to the differences in their isoelectric points; adsorption for acid orange 8 is maximized at pH 2 while pH 12 is optimal for methylene blue. In general, the adsorption behavior of both mesoporous carbon and activated carbon for these dyes shows a similar trend. Mesoporous carbon shows a higher adsorption capacity than activated carbon for all tested dyes, clearly highlighting the advantage of mesoporous carbon over activated carbon in removing organic pollutants.

The main parameter of an adsorbent (including mesoporous carbons) to determine its adsorption capacity for an organic molecule (or organic pollutant) is the specific surface area. To investigate this, this research group also carried out experiments to link the measured BET surface area of adsorbents (see Section 2.3.2.3 in Chapter 2) to the adsorption capacity for dyes. The BET surface area of the mesoporous carbon was found to be 1783 m^2/g from analysis of nitrogen adsorption/desorption data. From dye adsorption data, the surface coverage of methylene blue on the same mesoporous carbon sample gives a surface area of 1600 m^2/g, which is only marginally lower than the BET surface area. This suggests that methylene blue molecules are able to access the mesoporous system of the mesoporous carbon sample and was adsorbed in both the primary and secondary mesopores of the adsorbent. When comparing the surface coverage for acid orange 8 and rhodamine B on the same mesoporous carbon sample, a lower coverage (only 581 m^2/g, a roughly 60% reduction) was recorded for rhodamine B due to its larger molecular size and therefore harder to diffuse into the pores of adsorbent.

Other examples include a study comparing the adsorption behavior of two types of mesoporous carbon: CMK-5 and CMK-3 using several dyes (reactive blue 19, acid red 57 and fuchsin basic) at various concentrations. The most remarkable difference in the structure between CMK-3 and CMK-5 is that CMK-3 consists of carbon rods in contrast to the nanopipes of CMK-5 (see Fig. 3.4). Although both CMK-3 and CMK-5 were templated from SBA-15, using a chemical vapor deposition (CVD) method with ferrocene as the carbon source gives better control in deposition. Hence the pores in the SBA-15 were coated instead of filled as in CMK-3, leading to the formation of a network of nanopipes (Lei et al., 2008). Generally, CMK-5 performed better than CMK-3 (582 mg/g vs 410 mg/g) while activated carbon samples (as standards) recorded a 360 mg/g capacity for reactive blue 19. This is mainly attributed to the hollow pipe structure of CMK-5.

The carbonaceous surface of CMK-3 can also be oxidized to carry various oxygenated organic groups, mainly carboxylic acids, which reduces the surface hydrophobicity of the mesoporous carbon materials. Also, this generates a slightly acidic surface which may preferentially adsorb basic, or positively charged, molecules (or pollutants). However, when an oxidized CMK-3 was tested for the adsorption of fuchsin basic in the same study, it only showed a marginal improvement (around 10%) in adsorption capacity when compared with unmodified CMK-3.

When compared with activated carbon standards, ordered mesoporous silicas (CMK materials) are still very expensive to produce with a rather complicated procedure. Therefore, in order to justify their practicality, effective regeneration, and recyclability, is a must for CMK materials. In a recycling study, the adsorbed dye molecules must be removed by an effective and easy method while the nature of the surface should not be altered. In this study, the recyclability of CMK-5 had been tested via regeneration by solvent extraction and thermal annealing. Regenerated using solvent extraction, CMK-5 give a reduced adsorption capacity of 317 (53.7% of original) and 295 mg/g (50% of original) in the second and third cycle, respectively. This is likely to be due to some strongly bound dye molecules that were not removed during the regeneration step. Also, harsher regeneration may cause structural damage, hence reducing adsorption capacity. However, if spent CMK-5 was regenerated with thermal annealing, an adsorption capacity of 447 mg/g (77%), and 421 mg/g (72%) were recorded for the second and third cycles, showing a marginal improvement compared with solvent extraction.

Based on these examples, CMK-type of mesoporous carbon materials may have shown improvement in the adsorption/removal of pollutants. The less-than-perfect recycling results suggest that using CMK materials on a large scale for this area of applications still has a long way to go. Unless the cost of production for mesoporous carbons reduces significantly, these materials may only be suitable for other high-end applications.

3.3.4.2 Ordered mesoporous carbons as supports for heterogeneous catalysts

Carbon materials have served as support materials for heterogeneous catalysts for decades already and they have been popular choices. For example, palladium on carbon (Pd/C), a selective hydrogenation/reduction catalyst, has been widely available in commercial outlets, and used in industrial processes. One obvious advantage of using carbon support is its hydrophobicity. Many organic synthesis reactions need an organic solvent for the solvation of reactants and products. A hydrophobic catalyst will help in avoiding phase transfer between "water" and "oil" phase, hence enhancing the activity (or the rate/kinetics) for certain types of reactions.

Generally, activated carbons, as well as other porous carbon materials, are widely used for this purpose because of their low costs and wide availability. However, the random porous structure of activated carbon suggests that the selectivity of the metal/carbon catalysts solely relies on the metal component. As we have seen in Chapter 2, the mesoporous structure (of mesoporous silicas) could help in the size selection of reactants and products. In theory, mesoporous carbon could have the same effect. Therefore, scientists have carried out studies on using mesoporous carbon (e.g., CMK-type) materials as catalyst supports.

Metal-on-carbon catalysts are widely studied and widely available. To prepare metal-on-carbon catalysts, a standard impregnation method can be used. Mesoporous carbon supports can be used in the same way as other carbon supports. Figure 3.8 illustrate a standard protocol for preparing metal on mesoporous carbon catalysts. In step 1, deposition of a metal precursor can be carried out by dissolving a metal source (usually nitrate) in a polar, volatile solvent, e.g., ethanol and suspending carbon supports (say CMK-3) in the solution. By evaporating the solvent slowly, the metal nitrate can be deposited on the carbon surface evenly. The deposited metal nitrate is then decomposed to metal oxide under a thermal decomposition process (step 2) to form metal oxide particles (e.g., $2M(NO_3)_2 \rightarrow 2MO + 4NO_2 + O_2$). Usually, metal nitrate is the preferred precursor because other salts are associated with problems. For example, metal halides (including chlorides) release halogens to the catalyst and can poison the catalysts easily. Metal sulfates also release sulfur which is also considered a catalyst poison. Other salts such as carbonates and phosphates are usually less suitable for this purpose due to low solubility.

Metal oxide particles are usually not mobile and can be used as a "stable form" for catalyst storage. However, the active form of many metal-on-carbon catalysts is in the form of zero-valent metal particles (e.g., Ni/C). Therefore, a final step of reduction is performed to generate active metal particles, usually just before the reaction takes place. Sometimes, this is called the "activation" step. Hydrogen is the preferred reductant for this step as the only by-product is water. In some cases, a passivation step using a small amount of O_2 (<5%) in helium (He) at low temperature (say room temperature) is performed to ensure a longer life of the catalyst. Noble metals (e.g., Pt, Pd, Au) could bypass the reduction step because their oxides are not particularly stable, their metal particles can form by direct decomposition of their salts.

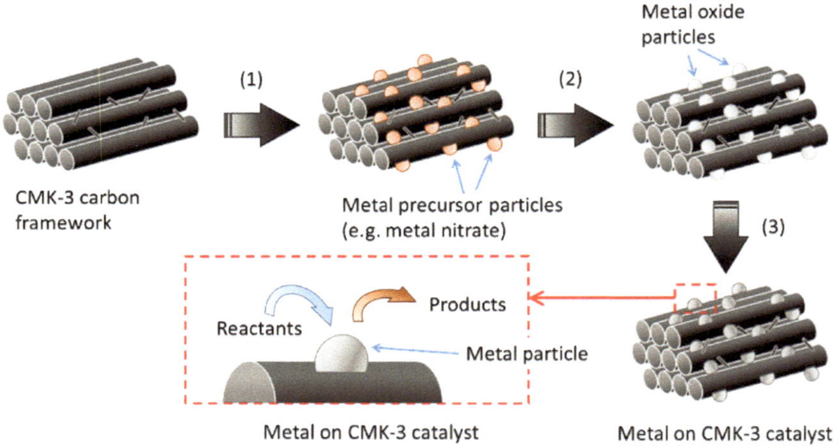

Figure 3.8 Illustration for the metal impregnation procedure for a metal/CMK-3 catalyst. Steps (1) metal precursor (usually nitrate) deposition, (2) thermal decomposition (converting nitrate to oxides), (3) temperature programmed reduction (reducing oxides to metals) using hydrogen gas. In some cases, "passivation" using 5% O_2 in He is required at the final stage to improve the stability of the catalyst.

There are plenty of examples in the literature using mesoporous carbon, in particular CMK-type, as support materials for catalysts, for a wide range of reaction types, with redox being the most commonly reported. In some cases, the chemistry of the CMK supports may need to be modified by doping with other elements such as nitrogen (e.g., Song et al., 2019). Table 3.1 shows several key examples using mesoporous carbon-based catalysts.

Table 3.1 Selected examples of the applications of functionalized ordered mesoporous carbon (OMC) in catalysis

Catalyst	Reaction type	Reaction	Ref
Sulfonated CMK-3 and CMK-8	Esterification	Esterification of free fatty acids in waste cooking oil	(Na et al., 2019)
Pd/OMC (soft template)	Redox/oxidation	Selective oxidation from benzyl alcohol to benzaldehyde	(Song et al., 2019)
Au/CMK-3	Redox/oxidation	Oxidation of Glucose to Gluconic Acid	(Qi et al., 2015)
Pd/CMK-3	Redox/hydrogenation	Hydrogenation of furfural to furfuryl alcohol	(Szewczyk et al., 2020)
Rh/CMK-8 and CMK-9	Syngas-to-alcohol (STA)	Higher alcohol synthesis from syngas	(Ha et al., 2015)
Co-Mo/CMK-3	Hydrodesulfurization	Conversion of thiophene to alkanes and alkenes	(Hussain & Ihm, 2009)

3.3.4.3 Electrical applications of ordered mesoporous carbons

Similar to other carbon materials including graphite and graphene, due to their conductive property, ordered mesoporous carbon can have great potential in electrical applications, used as electrical conductors or capacitors, or as components for electrodes in batteries or cells. Such applications are not conceivable for many other mesoporous materials (e.g., mesoporous silica or MCFs), due to their non-conductive nature. However, in some cases, modification on the ordered mesoporous carbon, e.g., by blending with a conductive polymer such as polyaniline or by doping with nitrogen, is required to achieve specific properties required for the desired applications. Table 3.2 shows several electrical applications of ordered mesoporous carbon and their derivatives.

Blending the ordered mesoporous carbon with a conductive polymer (e.g., polyaniline) to form a conductive composite can enhance the performance of the material. The electrochemical performance of the CMK-3 and polyaniline blended with nitrogen-doped ordered mesoporous carbon has been used in numerous examples in the literature (e.g., He et al., 2008; Dai et al., 2012). Another modification is doping, commonly with nitrogen. Doping with a foreign but compatible element (e.g., N) can generate structural "defects," which can further enhance the electrical properties. Another modification is to incorporate a foreign material at a micro-/nano-scopic level. This can offer extra properties to the mesoporous carbon. For example, metal oxide particles (Co_2O_3 and MnO_2) on mesoporous carbon surface enhances the electrical property of the carbon to become a supercapacitor (Zhao et al., 2015; Qu et al., 2014). On the other hand, supporting Pt nanoparticles on mesoporous carbon significantly enhances its catalytic property when used as a cathode for fuel cells (Kuppan & Selvam, 2012). Table 3.2 provides some examples in these areas found in the literature.

Table 3.2 Selected electrical applications for CMK-type mesoporous carbon materials

Electrical application	Type of capacitor/electrode materials	Ref
	As a capacitor	
• Supercapacitor	CMK-3 composite with cobalt oxide nanoparticles	(Zhao et al., 2015
• Supercapacitor	Polyaniline/nitrogen-doped OMC (CMK-3 type)	(Xie et al., 2018)
• Micro-capacitor	Mesoporous carbon thin film	(Garcia et al., 2019)
• Supercapacitor	MnO_2 in porous carbon coupled with N-OMC	(Qu et al., 2014)
	For electrode construction in batteries or cells	
• Cathode in lithium-sulfur batteries	Soft templated mesoporous carbon with sulfur composite	(Wang et al., 2015)
• Cathode for methanol fuel cell	Pt/CMK-3	(Kuppan & Selvam, 2012)

3.3.4.4 Doped Carbon Materials Derived from CMK Series

As mentioned, other elements can be incorporated into carbon materials to form compounds, offering a wide range of useful materials. The CMK family of mesoporous carbon can also be modified this way for expanding their breadth of applications. For example, mesoporous carbon nitride C_xN_y has been prepared using SBA-15 as the template with ethylenediamine and carbon tetrachloride as the nitrogen and carbon sources (Vinu et al., 2005). In particular, C_3N_4 would be useful with its semiconductive properties. Mesoporous C_3N_4 was prepared using a similar method to CMK-3 using SBA-15 as the hard template (Jun et al., 2009). Such C_3N_4 can also be doped with titanium (Ti) for the catalytic properties and the same research group also developed similar materials using other silica templates (e.g., silica spheres, ST60) to form different morphologies of C_3N_4 (Fischer et al., 2008).

Carbon nitrides (C_3N_4) have also shown good photocatalytic properties. By enhancing their surface area, ordered mesoporous C_3N_4 becomes superior for this application. For example, Mane et al. prepared mesoporous C_3N_4 using KIT-6 as the hard template and the material showed good activity in a photoactivated water-splitting reaction (Mane et al., 2017). The activity was found to be 267 $\mu mol\,h^{-1}$ of H_2 under a Xe lamp with a cut-off at 420 nm. Similar work has also been reported by Chen et al., who used SBA-15 as the hard template for preparing the mesoporous C_3N_4 material (Chen et al., 2009a). In a hydrogen evolution reaction from water under >595 nm radiation, a rate of ~85 $\mu mol\,h^{-1}$ of H_2 was recorded. Since the light sources used in these two examples were different, it would be difficult to make a direct comparison between the results being reported.

3.4 Non-Carboneous Mesoporous Materials via Hard Templating

In Chapter 2, we have seen the wide variety of ordered mesoporous silicas derived from the liquid crystal templating method. Their fascinating porous structures with high regularity and pore dimensions present a huge opportunity for materials science. Their uses range from catalysis, chromatography to biomedicine. The second generation, ordered mesoporous carbons (as seen in Section 3.3 here), also shows applications in completely different areas such as electronics, due to the conductive property. We should not forget that although silica is a versatile material, it is also fairly inert chemically and its chemical property is fairly limited. That is one reason why many uses of ordered mesoporous silicas heavily rely on functionalization (mainly post-synthesis), whether it is the incorporation of metal species (metal particles or metal substitution) or grafting organic groups (e.g., silanization).

The question is, "can we expand the panel of mesoporous materials towards functional materials without post-synthesis modifications?"

In the final sections of Chapter 2, we also introduced some mesoporous inorganic oxides templated using liquid crystals, or more appropriately micelles, since these mesoporous oxides are less ordered and the formation of a liquid crystal phase is not confirmed. Despite these successful examples, development in this research direction did not blossom. This is mainly because these mesoporous oxides are difficult to prepare and they are lacking mechanical strength. For example, Liu has reported the formation of mesostructured vanadium oxide (V_2O_5) via a soft template using deodecylamine as the structural directing reagent (Liu et al., 1997). However, the materials collapsed upon calcination at 500°C or solvent extraction of the surfactant template using HCl/ethanol. For many applications, these two obstacles must be overcome using different preparation methods.

One way to mitigate this is to use a hard template instead. When using soft templates, self-assembly is critical for a final material to have a long-range order. A hard template does not require the self-assembly of micelles as soft templates do. Therefore, the mesoporous materials prepared would have a stable long-range ordered porous structure. A similar method to the CMK series is therefore adapted for the synthesis of other inorganic materials using mesoporous silicas as hard templates.

3.4.1 Mesoporous Metal Oxides via Hard Templating

Similar to the synthesis of the CMK family of mesoporous porous carbons, we have to choose a silica template for mesoporous oxide synthesis to achieve the required structure for the desired applications. Among all common mesoporous silicas, the most attractive template candidates for casting mesoporous oxides are SBA-15, KIT-6, FDU-12 and SBA-16, due to their intrinsic 3-D porous structures. From this templating synthesis pathway, a number of crystalline mesoporous metals and transition metal oxides have been successfully prepared using mesoporous silica templates, e.g., Pt, Cr_2O_3, Co_3O_4, MnO_2 and V_2O_5 (Zhu et al, 2004;

Jiao et al., 2005). These crystalline materials can achieve a high surface area at 70–200 $m^2 g^{-1}$, comparing that of *ca.* 20 $m^2 g^{-1}$ for many bulk transition metal oxides; up to a 10-fold enhancement. At the same time, these materials still carry the basic properties of their bulk counterparts.

One key property of interest is again the electrical properties of these mesoporous transition metal oxides. Some mesoporous metal oxide electrodes exhibit high initial capacitance, with an excellent high-rate discharge performance and superior cyclic reversibility (i.e., allowing recharge cycles, which is essential for developing rechargeable batteries). With an enhanced surface area, the improvement in capacitance has been observed in many studies (Fang et al., 2021).

Among transition metal oxides, RuO_2, NiO or MnO_2, together with electronically conducting polymers such as polyaniline and polypyrrole, have been extensively studied for their use as capacitors. Some of these materials showed a specific capacitance higher than that of carbon materials being used in double-layer charge storage. This does attract the attention of material scientists (Simon and Gogotsi, 2008). One key issue being the largest obstacle is the low stability during cycling, leading to poor rechargeability if used in battery technology.

There are many examples of ordered mesoporous transition metal oxides used in various advanced applications. For instance, ordered mesoporous β-MnO_2, templated with KIT-6, has shown promising electrical properties. Lithium-ion batteries using ordered mesoporous β-MnO_2 electrodes have shown a high initial capacity and improved high-rate discharge performance with good cycling reversibility in an electrochemical measurement (Luo et al., 2006). Also, for use in lithium-ion batteries, ordered mesoporous Cr_2O_3 templated with KIT-6 was also tested as the cathode material, showing improved cycling stability when compared with bulk Cr_2O_3 (Dupont et al., 2008).

Moving away from electrical applications, the KIT-6 template was used for the preparation of ordered mesoporous iron oxides (Fe_3O_4, α-Fe_2O_3, and γ-Fe_2O_3) (Jiao et al., 2006). These iron oxides have unique magnetic properties and have found many applications in physics. Both the mesoporous Fe_3O_4 and γ-Fe_2O_3

materials showed magnetic freezing above 67°C (340 K) even though their wall thicknesses were around ~7 nm. Such a phenomenon was not observed with Fe_3O_4 or γ-Fe_2O_3 nanoparticles since these nanoparticles will lose their magnetic order at a particle size below 8 nm at a much lower temperature. This is very significant since many household appliances are operating at ambient temperature range, say 0–50°C, or 273–323 K, well covered by mesoporous iron oxides.

The catalytic properties of ordered mesoporous metal oxides were also studied. Many transition metal oxides are widely used as catalysts (e.g., Fe_3O_4 and V_2O_5); an increase in surface area due to the ordered mesoporous structure is a positive feature in catalysis. For example, KIT-6 templated CeO_2 was tested for CO oxidation (Shen et al., 2005). It showed enhanced activity compared with non-templated CeO_2, and the activity was further enhanced by the incorporation of CuO particles. More complex oxide catalysts can also be prepared using the hard-template method. For instance, mesoporous V_2O_5/WO_3 catalysts have been reported using SBA-15 as a template, with 4% V_2O_5 in the composite (Skliri et al., 2014). The mesoporous WO_3 catalyst has been shown active in the oxidation of 1-phenylethanol to acetophenone using tert-butyl hydroperoxide (t-BuOOH) as an oxidant. The addition of V_2O_5 in the structure has enhanced the activity compared with pure WO_3.

Tungsten oxide (WO_3) is also known to be a good photocatalyst, with the added advantage of activity under the visible spectrum. However, the problem of WO_3 as a photocatalyst, among other heavy metal oxides, is their low specific surface area. Using templating is a solution for this problem by increasing the surface. For example, Wen et al. prepared mesoporous WO_3 using SBA-15 as a template and tested the photocatalytic activity of the Pt-loaded mesoporous WO_3 samples (Wen et al., 2013). It was found that the mesoporous WO_3 have an enhanced specific surface area (152.7 m^2/g), a 15-fold improvement of that for bulk WO_3 (9.8 m^2/g). In a catalytic assay using methylene blue as a substrate under visible light ($\lambda > 400$ nm), mesoporous Pt-WO_3 also showed a 10-fold increase in degradation activity over bulk Pt-WO_3. Under the same conditions, commercial photocatalyst

TiO_2 (P25) showed virtually no activity. It is worth noting that the deposition of Pt particles can enhance the photocatalytic activity of WO_3 under visible radiation. In another example, mesoporous WO_3 templated using KIT-6 showed a similar structural effect, possessing a surface area of 151 m^2/g (Villa et al., 2015). In this work, the oxidation of methane CH_4 formed methanol CH_3OH using H_2O_2 under visible radiation. The mesoporous WO_3 activity was enhanced by the addition of Fe^{3+} as an electron scavenger. Use of Ag^+ as a scavenger led to a high selectivity towards CO_2 formation.

3.4.2 Mesoporous Metal Nitrides via Hard Templating

In addition to oxides, metal nitrides also caught the attention of scientists due to their potential in electrical applications. For example, vanadium nitride (VN) has shown its use as a supercapacitor. Similar to metal oxides, increasing the surface area of nitrides will open up many applications by enhancing the material's performance. In 1999, Winkler et al. became the first to report the synthesis of metal nitride, GaN in this pioneering study, inside the channels of mesoporous silica templates in 1999 (Winkler et al., 1999). In this study, an organometallic precursor triazido (trimethylamine) gallium cluster was deposited inside the mesopores of the MCM-41 silica template. GaN was formed by decomposing the precursor at 500°C under purging ammonia. Using a similar strategy, titanium nitrides (TiN) were prepared inside the mesopores of SBA-15 using $Ti(NMe_2)_4$ as the nitride precursor (Hsueh et al., 2005). Again decomposition occurred at an elevated temperature, 700°C in this study, and also under an ammonia atmosphere. The same research went further to prepare gallium nitride (GaN) (Hsueh et al., 2005) and indium nitride (InN) silica composites (Chang and Huang, 2008) using SBA-15 as a template, with $GaCl_3$ or $In(NO)_3$, respectively, as precursors. In these two studies, the removal of the silica templates was attempted using HF or NaOH. However, in both studies, the desired mesostructure was not retained once the silica template was removed, and a "nano-rod" of the size of the SBA-15 mesopores was formed. This suggested that the interconnecting micropores between the SBA-15 mesopores were not fully filled with precursors or the nitrides were too weak mechanically to withstand the harsh template removal regime.

To achieve truly mesoporous metal nitrides using mesoporous silica as hard templates, two synthesis pathways have been reported as shown in Fig. 3.9. To start, the metal oxide was formed inside the mesopores of the silica template, resulting in a metal oxide/silica composite with a mesoscale structure. In Route A (blue route), the silica template was removed to form a mesostructured metal oxide, followed by ammonia nitridation to form mesoporous nitrides. In contrast, in Route B, the ammonia nitridation step was carried out to the metal oxide/silica mesostructured composite before removing the silica template.

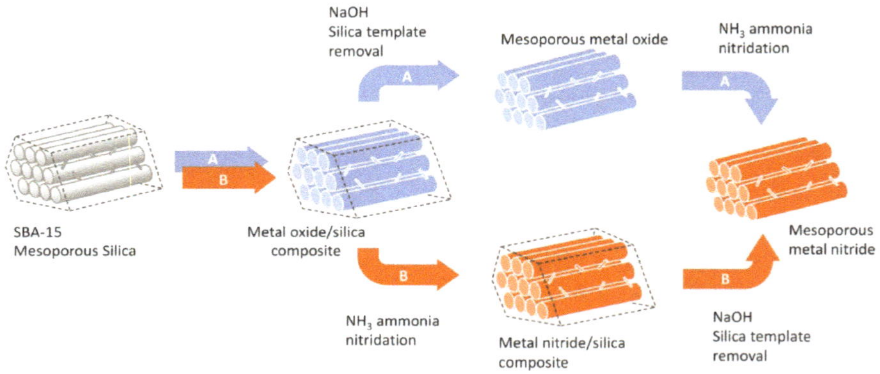

Figure 3.9 The illustration depicts two synthesis pathways (Route A, blue arrows, and Route B, orange arrows) for mesoporous metal nitrides using mesoporous silica as the hard templates.

Which route to use depends on the stability of the intermediates. For example, in Route A, the mesostructure of the metal oxides needed to be stable to withstand the harsh conditions of ammonia nitridation treatment without the support of the silica templates. Take mesoporous CoN as an example, it was successfully prepared with mesoporous Co_3O_4 formed form via nanocasting using silica SBA-15 as a hard template, followed by ammonia nitridation to form mesoporous CoN (Shi et al., 2008). This mesoporous CoN showed a high BET surface area (89 $m^2\ g^{-1}$) and a high pore volume (0.2 $m^3\ g^{-1}$), comparable to its oxide intermediate. However, the final product is not stable at high temperatures as the mesostructure collapsed at a temperature above 375°C. This structural collapse is thought to be linked to the shrinkage that occurs during the transformation from oxide to nitride.

In Route B, ammonia nitridation took place when mesostructured metal oxide was still within the mesopores of a silica template. This could circumvent the collapse of the nitrides during ammonia nitridation. However, these nitrides still have to survive the harsh corrosive conditions during the silica template removal process with either NaOH or HF solutions. For example, a mesostructured copper (I) nitride (Cu_3N) was prepared using carbon-lined SBA-15 as the hard template with copper acetylacetonate as a copper precursor (Zhang et al., 2009). A Cu-C-SBA-15 composite was obtained and finally submitted for ammonia nitridation. Since Cu_3N can be dissolved in HF and basic solution, there was no attempt to remove the silica template and a mesostructured Cu_3N/silica composite was the final product. Shi et al. have used this route to prepare ordered mesoporous CrN and CoN by ammonia nitridation of Cr_2O_3 and Co_2O_3 to CrN and CoN, followed by template removal using NaOH (Shi et al., 2008). One interesting finding was that the nitridation temperature needed to be carefully chosen for optimal stability and this temperature varied greatly depending on the formulation of the nitride. In this study, nitridation to form CrN is about 500°C higher than that of CoN. Such temperature could be crucial for the stability of any delicate mesoporous materials.

3.4.3 Application of Mesoporous Metal Nitrides in Electronics

With their electrical properties, transition metal nitrides are promising candidates as supercapacitor materials, which attract much attention in scientific communities (Choi et al., 2006a; Liu et al., 1998). In the early stage of development, these transition metal nitrides were found to exhibit pseudocapacitive behavior in aqueous electrolytes (Chen et al., 2009b). Since transition metal nitrides show a high electronic conductivity (Glushenkov et al., 2010) and high stability in aqueous electrolytes (Liu et al.,

1998), if they can be made to have high surface areas (Pande et al., 2012), performance in electronics can be improved.

Take VN as an example; it is one of the most studied materials for supercapacitor applications (Choi et al., 2006b) due to its high capacitance, up to 1340 F g^{-1}. The main obstacle is that there are still knowledge gaps on how the charge is stored in nitride materials. A 2-step mechanism was suggested for the charge storage system of Mo_2N in dilute sulfuric acid. The first is the reduction of Mo with protons followed by an oxy/hydroxyl redox couple reaction on the hydrolyzed Mo (Choi et al., 2006).

The high capacitance recorded from VN is thought to link to a thin layer of VO_x or oxynitrides formed on the VN surface, leading to a high pseudocapacitance. However, VN materials are not perfect as a noticeable decrease in capacitance was observed when a high "sweep rate" was performed during a cyclic voltammetric measurement. This phenomenon was thought to be linked to the microscopic structure of VN, which is usually built upon weakly bound VN nanoparticles. Such structural defects affect electrical conductivity. One proposed solution lies in reducing the dimension of the particles to improve the charge/discharge rate.

With increased surface area and porosity, the mesoporous structures of a material can be considered for use as a material for constructing electrodes. Mesoporous structures provide an increased surface area to increase the total charge available and enhance the electrolyte accessibility and diffusivity. However, research development in this area is surprisingly slower than anticipated. Only recently, nickel nitride and nickel-iron nitride were prepared with Route A using KIT-6 as the hard template (see Fig. 3.9) and tested for their electrocatalytic performance in oxygen evolution reaction (Saad et al., 2020). This is possibly due to the technical difficulties in a rather complicated synthesis procedure and the practicality of scaling up even if the materials are proven to have enhanced properties.

3.4.4 Other Materials Prepared with Hard-Template Methods

Using the same principle, metal phosphides with ordered structures can also be made. Metal phosphides also display good electrical properties that lead to a wide range of electrical and electronic applications. For example, cobalt and nickel phosphides, as well as cobalt-nickel bimetallic phosphides, were prepared using KIT-6 as the hard template (Fu et al., 2016). These metal phosphides were used to construct the work electrode, together with RuO_2, and tested for oxygen evolution reaction. For applications in the same area, perovskite oxides were also prepared. Wang et al. demonstrated the preparation of a series of $LaFe_xCo_{1-x}O_3$ perovskite oxides also with KIT-6 as the hard template (Wang et al., 2013). These perovskites were also tested as electrode materials, but essentially without the addition of RuO_2, and their electrocatalytic activity for oxygen evolution reaction was assessed.

Apart from inorganics, porous organic materials can also be made via the hard-template pathway, using ordered mesoporous silica as the hard templates. Very often these organic materials are synthetic polymers. Generally, it is a three-step process; (1) loading the monomer molecules inside the mesoporous of the template, (2) polymerization, and (3) template removal with NaOH or HF. This process is depicted in Fig. 3.10. Three criteria must be met for the successful synthesis of mesoporous polymers using this hard-template route. First, the polymerization process must be able to take place inside the mesopores. Therefore, the monomers must be small enough to be deposited inside the mesopores. Then, the formed polymer itself must be chemically resistant to the harsh process of silica removal using either HF or NaOH. This rules out many common polymers such as polyesters and polycarbonates as these will depolymerize in a basic medium. Finally, the polymer material must be rigid enough to retain the mesoporous structure after the removal of the silica template.

Initially, this method was applied to make nanostructures of common polymers such as polyethylene and polypropylene. In order to ensure the polymerization took place inside the mesopores of the templates, polymerization catalysts such as Cp_2TiCl_2 were grafted onto the walls of the mesopores. However, these materials formed as nanofibers instead of ordered mesostructures. More importantly, polyethylene and polypropylene are not highly sought-after materials with few chemical properties that catch scientists' attention (Dong et al., 2005; Dong et al., 2006; Ye et al.,

2003). Therefore, very little development from this direction has been seen.

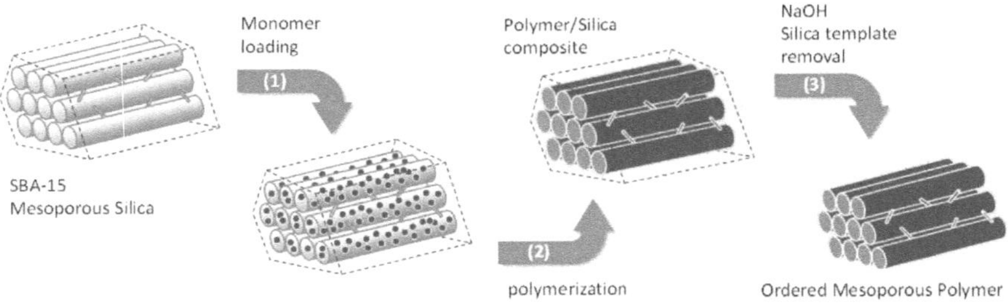

Figure 3.10 Illustration depicts the three-step process for the synthesis of mesoporous polymers using the hard-template route.

There are other nanofibers of more intriguing polymers made from this method. The first truly polymer replicate of an ordered mesoporous system was reported in 2001 when poly-divinylbenzene (polyDVB) systems were prepared using MCM-48 and SBA-15 as hard templates (Kim et al., 2001). The mesostructured polyDVB systems survived after the removal of the silica templates and high surface areas (>400 m^2 g^{-1}) were also recorded. Other similar systems also include a polypyrrole/PMMA composite templated with SBA-15 (Jang et al., 2001).

3.5 Third Generation of Mesoporous Materials

Although mesoporous metal oxides and nitrides can be made from hard templates similar to the synthesis pathways as CMK mesoporous carbon series, the structure of these templated oxides and nitrides are still not an exact replicate of SBA-15 or KIT-6, which have unique 2-D and 3-D porous structure. To achieve this, the template needs to be the CMK carbons (as shown in Fig. 3.11), to generate a "third generation" of mesoporous materials. One clear advantage of using mesoporous carbon is that the template can be removed by simple combustion. Very little carbon residue will remain in the final product. Also, the procedure is less clumsy when compared with the removal of SiO_2 using either HF or NaOH, both are considerably corrosive. The caveat is that such a procedure will have to be performed during the preparation of the CMK templates.

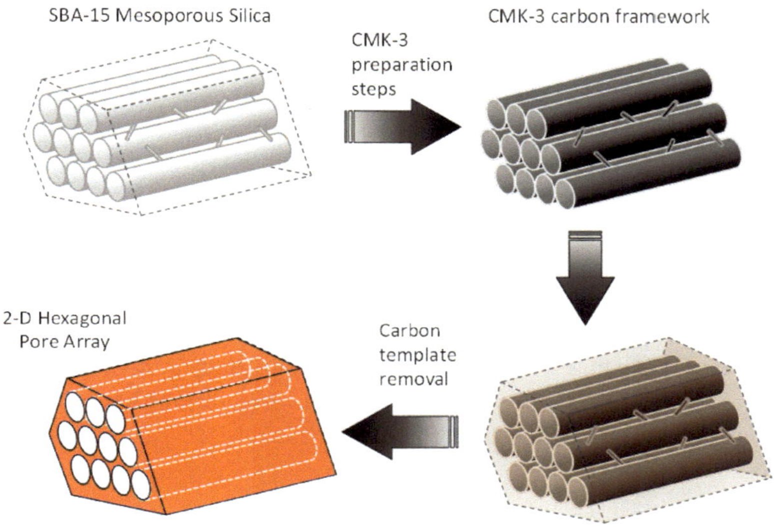

Figure 3.11 Synthesis pathways for the third generation of ordered mesoporous solid from hard templating method.

One early example that used this method to prepare the third generation of mesoporous materials was reported by Roggenbuck and Tiemann, who prepared mesoporous MgO using CMK-3 as the hard template (Roggenbuck and Tiemann, 2005). This material showed a remarkable 306 $m^2\ g^{-1}$, which is very high for MgO, and tested for CO_2 adsorption/desorption using TPD. Since the materials went through high-temperature calcination upon removal of the CMK-3 template, the thermal stability of this mesoporous MgO is unquestionable. However, also due to the thermal treatment, the mesostructured (d-spacing) has shown a shrinkage of about 10%. Another example of a third generation of mesoporous materials was reported by Jun et al. as mesoporous titanium nitride (TiN)-carbon composite was prepared using mesoporous C_3N_4 (prepared from SBA-15) as the hard template (Jun et al., 2009).

Although the third generation of ordered mesoporous solids can successfully be made as true replicates of the sophisticated

SBA or KIT series structure, only limited research reports can be found in the literature. This method is by far less popular than other methods such as soft templating. One reason is that this multi-step method is just too laborious for many researchers. Such a complicated procedure has two implications; first, the third-generation mesoporous materials are not economical to manufacture and scaling up production is unlikely to achieve. Moreover, there is still no proof that these exact replicate structures are superior in properties when compared with other mesoporous structures of the same material, e.g., 2nd generation. Consequently, this third generation still stays in the research stage. In theory, though, the fourth and higher generations of mesoporous materials can be made using this route but the meso-scaled structure will be gradually diminished, generation after generation. Also the complexity of the preparation becomes not worth to be considered.

3.6 Concluding Remarks

The wide range of mesoporous materials, such as carbon, metal oxides, nitrides, and phosphides, and complex structures, such as perovskites, have been introduced in this chapter, and the applications of mesoporous materials have been extended to electrical and electronics, as well as photocatalysis. The synthesis methodologies presented here also allowed scientists to develop materials that are simply too fragile to form using soft templates. However, the pattern of development is still derived from a bulk material and then enhancing their physical properties, most likely to be increasing the surface area, and maybe the porosity. Although not all bulk materials can be improved this way, it does give a good research direction to scientists to improve the developed materials. Still, similar to many newly developed materials, the obstacle to development is on scaling up production. Without a viable route for scaling up economically, these materials are simply too expensive to be manufactured and used in daily applications.

References

Agustian, J., and Hermida, L. (2019). Mesostructured cellular foam MCF-(9.2T-3D) silica as support for free αamylase in liquefaction of tapioca starch, *Mater. Sci. Eng.*, **509**, p. 012059.

Beck, J. S., Vartuli, J. C., Roth, W. J., Leonowicz, M. E., Kresge, C. T., Schmitt, K. D., Chu, C. T. W., Olson, D. H., Sheppard, E. W., McCullen, S. B., Higgins, J. B., and Schlenker, J. L. (1992). A new family of mesoporous molecular sieves prepared with liquid crystal templates, *J. Am. Chem. Soc.* **114**, pp. 10834–10843.

Böke, N., Godongwana, Z. G., and Petrik, L. F. (2013). Synthesis of an ordered mesoporous carbon with graphitic characteristics and its application for dye adsorption, *J. Porous Mater.*, **20**, pp. 1153–1161.

Chang, S. C., and Huang, M. H. (2008). Formation of indium nitride nanorods within mesoporous silica SBA-15, *Inorg. Chem.*, **47**, pp. 3135–3139.

Chen, X., Jun, Y.-S., Takanabe, K., Maeda, K., Domen, K., Fu, X., Antonietti, M., and Wang, X. (2009a). Ordered mesoporous SBA-15 type graphitic carbon nitride: a semiconductor host structure for photocatalytic hydrogen evolution-with visible light, *Chem. Mater.*, **21**, pp. 4093–4095.

Chen, H., Shu, D., He, C., and Nan, J. (2009b). Study on the electrochemical behaviour of vanadium nitride as a promising supercapacitor material, *J. Phys. Chem. Solids*, **70**, pp. 495–500.

Choi, D., and Kumpta, P. N. (2006a). Nanocrystalline TiN derived by a two-step halide approach for electrochemical capacitors, *J. Electrochem. Soc.*, **153**, pp. 2298–2303.

Choi, D., Blomgreen, G. E., and Kumta, P.N. (2006b). Fast and reversible surface redox reaction in nanocrystalline vanadium nitride super-capacitors, *Adv. Mater.*, **18**, pp. 1178–1182.

Dai, H., Lin, Y., Xu, G., Gong, L. Yang, C., Ma, X., and Chen, G., (2012). Cathodic electrochemiluminescence of luminol using polyaniline/ordered mesoporous carbon (CMK-3) hybrid modified electrode for signal amplification, *Electrochim. Acta*, **78**, pp. 508–514.

Dong, X. C., Wang, L., Jiang, G. H., Zhao, Z. R., Sun, T. X., Yu, H. J., and Wang, W. Q. (2005). MCM-41 and SBA-15 supported Cp_2ZrCl_2 catalysts for the preparation of nano-polyethylene fibres via *in situ* ethylene extrusion polymerization, *J. Mol. Catal. A: Chem.*, **240**, pp. 239–244.

Dong, X. C., Wang, L., Wang, J. J., Zhou, J. F., and Sun, T. X. (2006). Evidence of nanoconfinement effects of MCM-41 on propylene polymerization catalyzed by MCM-41 supported metallocene catalyst in the presence and absence of β-cyclodextrin, *J. Phys. Chem. B*, **110**, pp. 9100–9104.

Dupont, L., Laruelle, S., Grugeon, S., Dickinson, C., Zhou, W., and Tarascon, J.-M. (2008). Mesoporous Cr_2O_3 as negative electrode in lithium batteries: TEM study of the texture effect on the polymeric layer formation. *J. Power Sources*, **175**, pp. 502–509.

Eftekhari, A., and Fan, Z. (2017). Ordered mesoporous carbon and its applications for electrochemical energy storage and conversion. *Mater. Chem. Front.*, **1**, pp. 1001–1027.

Fang, Y., Zhang, Q., and Cui, L. (2021). Recent progress of mesoporous materials for high performance supercapacitors, *Microporous Mesoporous Mater.*, **314**, p. 110870.

Feng, X., Hu, G., Hu, X., Xie, G., Xie, Y., Lu, J., and Luo, M. (2013). Tetra-ethylenepentamine-modified siliceous mesocellular foam (MCF) for CO_2 capture, *Ind. Eng. Chem. Res.*, **52**, pp. 4221–4228.

Fischer, A., Jun, T.-S., Thomas, A., and Markus Antonietti (2008). Synthesis of high-surface-area TiN/carbon composite materials with hierarchical porosity via "reactive templating", *Chem. Mater.* **20**, pp. 7383–7389.

Fu, S., Zhu, C., Song, J., Engelhard, M. H., Li, X., Du, D., and Lin, Y. (2016). Highly ordered mesoporous bimetallic phosphides as efficient oxygen evolution electrocatalysts. *ACS Energy Lett.*, **1**, pp. 4792–4796.

Gao, F., Huang, J., Sun, H., Hu, J., Wang, M., Mi, J., and Wu, C. (2019). CO_2 capture using mesocellular siliceous foam (MCF)-supported CaO. *J. Energy Inst.*, **92**, pp. 1591–1598.

Garcia, S., Lozano-Castello, D., Morallon, E., and Cazorla-Amoros, D. (2019). Silica-templated ordered mesoporous carbon thin films as electrodes for micro-capacitors, *J. Mater. Chem. A*, **4**, pp. 4570–4579.

Glushenkov, A. M., Hulicova-Jurcakova, D., Llewellyn, D., Lu, G. Q., and Chen, Y. (2010). Structure and capacitive properties of porous nanocrystalline VN prepared by temperature-programmed ammonia reduction of V_2O_5, *Chem. Mater.*, **22**, pp. 914–921.

Han, Y.-J., Stucky, G. D., and Butler, A. (1999). Mesoporous silicate sequestration and release of proteins, *J. Am. Chem. Soc.*, **121**, pp. 9897–9898.

Han, Y., Lee, S. S., and Ying, J. Y. (2007). Spherical siliceous mesocellular foam particles for high-speed size exclusion chromatography, *Chem. Mater.*, **19**, pp. 2292–2298.

He, Y., Cheng, Q., Pavlinek, V., Li, C., and Saha, P. (2008). Synthesis and structural characterization of polyaniline/mesoporous carbon nanocomposite, *Int. J. Polym. Anal. Charact.*, **13**, pp. 25–36.

Hsueh, H. S., Yang, C. T., Zink, J. I., and Huang, M. H. (2005). Formation of titanium nitride nanoparticles within mesoporous silica, SBA-15, *J. Phys. Chem. B*, **109**, pp. 4404–4409.

Hussain, M., and Ihm S.-K., (2009). Synthesis, Characterization, and hydrodesulfurization activity of new mesoporous carbon supported transition metal sulfide catalysts, *Ind. Eng. Chem. Res.*, **48**, pp. 698–707.

Jang, J. S., Lim, B., Lee, J., and Hyeon, T. (2001). Fabrication of a novel polypyrrole/poly(methyl methacrylate) coaxial nanocable using mesoporous silica as a nanoreactor, *Chem. Commun.* pp. 83–84.

Janus, R., Natkański, P., Wądrzyk, M., Lewandowski, M., Michalik, M., and Kuśtrowski, P. (2022). Deposition of poly(furfuryl alcohol) in mesoporous silica template controlled by solvent polarity: a cornerstone of facile and versatile synthesis of high-quality CMK-type carbon replicas. Nanocasting of SBA-15, SBA-16, and KIT-6. *Carbon*, **195**, pp. 292–307.

Jiao, K., Zhang, B., Yue, B., Ren, Y., Liu, S. X., Yan, S. R., Dickinson, C., Zhou, W., and He, H. (2005). Growth of porous single-crystal Cr_2O_3 in a 3-D mesopore system, *Chem. Commun.*, pp. 5618–5620.

Jiao, F., Jumas, J.-C., Womes, M., Chadwick, A. V., Harrison, A., and Bruce, P. G. (2006). Synthesis of ordered mesoporous Fe_3O_4 and γ-Fe_2O_3 with crystalline walls using post-template reduction/oxidation, *J. Am. Chem. Soc.*, **128**, pp. 12905–12909.

Joo, S. H., Jun, S., and Ryoo, R. (2001). Synthesis of ordered mesoporous carbon molecular sieves CMK-1, *Microporous Mesoporous Mater.*, **44–45**, pp. 153–158.

Jorio, A., Dresselhaus, M. S., Saito, R., and Dresselhaus, G. (2011). The sp^2 Nanocarbons: Prototypes for Nanoscience and Nanotechnology, in *Raman Spectroscopy in Graphene Related Systems*, Chapter 1 (WILEY-VCH Verlag, Weinheim, Germany), pp. 1–15.

Jun, Y.-S., Hong, W. H., Antonietti, M., and Thomas, A. (2009). Mesoporous, 2D hexagonal carbon nitride and titanium nitride/carbon composites, *Adv. Mater.*, **21**, pp. 4270–4274.

Jung, D., and Hartmann, M. (2010). Oxidation of indole with CPO and GO_x immobilized on mesoporous molecular sieves, *Catal. Today*, **157**, pp. 378–383.

Kannan, K., and Jasra, R. V. (2011). Improved catalytic hydrolysis of carboxy methyl cellulose using cellulase immobilized on functionalized meso cellular foam, *J. Porous Mater.*, **18**, pp. 409–416.

Kim, J. Y., Yoon, S. B., Kooli, F., and Yu, J. S. (2001). Synthesis of highly ordered mesoporous polymer networks, *J. Mater. Chem.*, **11**, pp. 2912–2914.

Kim, T.-W., Kim, M.-J., Chae, H.-J., Ha K.-S., and Kim, C. U., (2015). Ordered mesoporous carbon supported uniform rhodium nanoparticles as catalysts for higher alcohol synthesis from syngas, *Fuel*, **160**, pp. 393–403.

Kuppan, B., and Selvam P. (2012). Platinum-supported mesoporous carbon (Pt/CMK-3) as anodic catalyst for direct methanol fuel cell applications: the effect of preparation and deposition methods. *Prog. Nat. Sci. Mater. Int.* **22**, pp. 616–623.

Lei, Z., Bai, S., Xiao, Y., Dang, L., An, L., Zhang, G., and Xu, Q. (2008). CMK-5 mesoporous carbon synthesized via chemical vapor deposition of ferrocene as catalyst support for methanol oxidation, *J. Phys. Chem. C*, **112**, pp. 722–731.

Liu, P., Moudrakovski, I. L., Liu, J., and Sayari, A. (1997). Mesostructured vanadium oxide containing dodecylamine, *Chem. Mater.* **9**, pp. 2513–2520.

Liu, T. C., Pell, W. G., Conway, B. E., and Roberson, S. L. (1998). Behavior of molybdenum nitrides as materials for electrochemical capacitors comparison with ruthenium oxide, *J. Electrochem. Soc.*, **145**, pp. 1882–188.

Liu, T., Wang, K., Jiang, M., and Wan, L. (2019). A drug release model constructed by factorial design to investigate the interaction between host mesoporous silica carriers and drug molecules, *AAPS Pharm. Sci. Tech.*, **20**, p. 126.

Luo, J. Y., Zhang, J. J., and Xia Y. Y. (2006). Highly electrochemical reaction of lithium in the ordered mesoporous beta-MnO_2, *Chem. Mater.*, **18**, pp. 5618–5623.

Mane, G. P., Talapaneni, S. N., Lakhi, K. S., Ilbeygi, H., Ravon, U., Al-Bahily, K., Mori, T., Park, D. H., and Vinu, A. (2017). Highly ordered nitrogen-rich mesoporous carbon nitrides and their superior performance for sensing and photocatalytic hydrogen generation, *Angew. Chem. Int. Ed.*, **56**, pp. 8481–8485.

Na, S., Minhua, Z., Xiuqin, D., and Lingtao, W. (2019). Preparation of sulfonated ordered mesoporous carbon catalyst and its catalytic performance for esterification of free fatty acids in waste cooking oils, *RSC Adv.*, **9**, pp. 15941–15948.

Nanaki, S., Tseklima, M., Christodoulou, E., Triantafyllidis, K., Kostoglou, M., and Bikiaris, D. N. (2017). Thiolated chitosan masked polymeric microspheres with incorporated mesocellular silica foam (MCF) for intranasal delivery of paliperidone, *Polymers (Basel)*, **9**, 617.

Pande, P., Rsmussen, P. G., and Thompson, L. T. (2012). Charge storage on nanostructured early transition metal nitrides and carbides, *J. Power Sources*, **207**, 212–215.

Qi, P., Chen, S., Chen, J., Zheng, J., Zheng, X., and Yuan, Y. (2015). Catalysis and reactivation of ordered mesoporous carbon-supported gold nanoparticles for the base-free oxidation of glucose to gluconic acid, *ACS Catal.*, **5**, pp. 2659–2670.

Qu, D., Wen, J., Liu, D., Xie, Z., Zhang, X., Zheng, D., Lei, J., Zhong, W., Tang, H., Xiao, L., and Qu, D. (2014). Hydrogen ion supercapacitor: a new hybrid configuration of highly dispersed MnO_2 in porous carbon coupled with nitrogen-doped highly ordered mesoporous carbon with enhanced H-insertion, *ACS Appl. Mater. Interfaces*, **6**, pp. 22687–22694.

Roggenbuck, J., and Tiemann, M. (2005). Ordered mesoporous magnesium oxide with high thermal stability synthesized by exotemplating using CMK-3 carbon, *J. Am. Chem. Soc.*, **127**, pp. 1096–1097.

Ryoo, R., Joo, S. H., Kruk, M., and Jaroniec, M. (2001). Ordered mesoporous carbons, *Adv. Mater.*, **13**, pp. 677–681.

Saad, A., Shen, H., Cheng, Z., Arbi, R., Guo, B., Hui, L. S., Liang, K., Liu, S., Attfield, J. P., Turak, A., Wang, J., and Yang, M. (2020). Mesoporous ternary nitrides of earth-abundant metals as oxygen evolution electrocatalyst. *Nano-Micro Lett.*, **12**, p. 79.

Schmidt-Winkel, P., Lukens Jr., W. W., Zhao, D., Yang, P., Chmelka, B. F., and Stucky, G. D. (1999). Mesocellular siliceous foams with uniformly sized cells and windows, *J. Am. Chem. Soc.*, **121**, pp. 254–255.

Shen, W., Dong, X., Zhu, Y., Chen, H., and Shi L. (2005). Mesoporous CeO_2 and CuO-loaded mesoporous CeO_2: synthesis, characterization, and CO catalytic oxidation property. *Microporous Mesoporous Mater.*, **85**, pp. 157–162.

Shi, Y., Wan, Y., Zhang, R., and Zhao, D. Y. (2008). Synthesis of self-supported ordered mesoporous cobalt and chromium nitrides, *Adv. Funct. Mater.*, **18**, pp. 2436–2443.

Simon, P., and Gogotsi, Y. (2008). Materials for electrochemical capacitors, *Nat. Mater.*, **7**, pp. 845–854.

Skliri, E., Lykakisb, I. N., Gerasimos, S., and Armatas, G. S. (2014). Ordered mesoporous V_2O_5/WO_3 composite catalysts for efficient oxidation of aryl alcohols, *RSC Adv.*, **4**, pp. 46170–46178.

Sneddon, G., Greenaway, A., and Yiu, H. H. P. (2014). The potential applications of nanoporous materials for the adsorption, separation, and catalytic conversion of carbon dioxide, *Adv. Energy Mater.*, **4**, p. 1301873.

Sneddon, G., Ganin, A. Y., and Yiu, H. H. P. (2015). Sustainable CO_2 adsorbents prepared by coating chitosan onto mesoporous silicas for large-scale carbon capture technology, *Energy Technol.*, **3**, pp. 249–258.

Sneddon, G., McGlynn, J. C., Neumann, M., Aydın, H. M., Yiu, H. H., and Ganin, A. Y. (2017). Aminated poly(vinyl chloride) solid state adsorbents with hydrophobic function for post-combustion CO_2 capture, *J. Mater. Chem. A*, **5**, pp. 11864–11872.

Song, H., Liu, Z., Gai, H., Wang, Y., Qiao, L., Zhong, C., Yin, X., and Xiao, M. (2019). Nitrogen-dopped ordered mesoporous carbon anchored Pd nanoparticles for solvent free selective oxidation of benzyl alcohol to benzaldehyde by using O_2, *Front. Chem.*, **7**, p. 458.

Sun, S., Zhang, X., Han, Q., Wan, W., and Ding, M. (2016). Preparation and retention mechanism exploration of mesostructured cellular foam silica as stationary phase for high performance liquid chromatography, *Talanta*, **149**, pp. 187–193.

Szewczyk, I., Kosydar, R., Natkański, P., Duraczyńska, D., Gurgul, J., Kuśtrowski, P., and Drelinkiewicz, A. (2020). Effect of the type of siliceous template and carbon precursor on physicochemical and catalytic properties of mesoporous nanostructured carbon-palladium systems, *J. Porous Mater.*, **27**, pp. 1287–1308.

Villa, K., Murcia-López, S., Andreua, T., and Morante, J. R. (2015). Mesoporous WO_3 photocatalyst for the partial oxidation of methane to methanol using electron scavengers, *Appl. Catal. B: Environ.*, **163**, pp. 150–155.

Vinu, A., Ariga, K., Mori, T., Nakanishi, T., Hishita, S., Golberg, D., and Bando, Y. (2005). Preparation and characterization of well-ordered hexagonal mesoporous carbon nitride, *Adv. Mater.*, **17**, pp. 1648–1652.

Wang, Y., Cui, X., Li, Y., Chen, L., Shu, Z., Chen, H., and Shi, J. (2013). High surface area mesoporous $LaFe_xCo_{1-x}O_3$ oxides: synthesis and electrocatalytic property for oxygen reduction, *Dalton Trans.*, **42**, pp. 9448–9452.

Wang, H., Zhang, C., Chen, Z., Liu, H., and Guo Z. (2015). Large-scale synthesis of ordered mesoporous carbon fiber and its application as cathode material for lithium-sulfur batteries, *Carbon*, **81**, pp. 782–787.

Wen, Z., Wu, W., Liu, Z., Zhang, H., Li, J., and Chen, J., (2013). Ultrahigh-efficiency photocatalysts based on mesoporous Pt–WO_3 nanohybrids, *Phys. Chem. Chem. Phys.,* **15**, pp. 6773–6778.

Winkler, H., Birkner, A., Hagen, V., Wolf, I., Schmechel, R., Von Seggern, H., and Fischer, R. A. (1999). Quantum-confined gallium nitride in MCM-41, *Adv. Mater.*, **11**, pp. 1444–1448.

Xin, W., and Song, Y. (2015). Mesoporous carbons: recent advances in synthesis and typical applications, *RSC Adv.,* **5**, pp. 83239–83285.

Xie, K., Zhang, M., Yang, Y., Zhao, L., and Qi, W. (2018). Synthesis and supercapacitor performance of polyaniline/nitrogen-doped ordered mesoporous carbon composites. *Nanoscale Res. Lett.*, **13**, p. 163.

Yan, K., Sun, X., Ying, Cheng, W., Deng, Y., Ma, Z., Zhao, Y., Wang, X., S. Pan, L., and Shi, Y. (2020). Ultrafast microwave synthesis of rambutan-like CMK-3/carbon nanotubes nanocomposites for high-performance supercapacitor electrode materials, *Sci. Rep.,* **10**, p. 6227.

Ye, Z. B., Zhu, S. P., Wang, W. J., Alsyouri, H., and Lin, Y. S. (2003). Morphological and mechanical properties of nascent polyethylene fibers produced via ethylene extrusion polymerization with a metallocene catalyst supported on MCM-41 particles, *J. Polym. Sci. Part B: Polym. Phys.,* **41**, pp. 2433–2443.

Yoshida, H., Oehlenschlaeger, S., Minami, Y., and Terashima, M. (2002). Adsorption of CO_2 on composites of strong and weak basic anion exchange resin and chitosan, *J. Chem. Eng. Jpn.,* **35**, pp. 32–39.

Zdarta, J., Feliczak-Guzik, A., Siwińska-Ciesielczyk, K., Nowak, I., and Jesionowski, T. (2020). Mesostructured cellular foam silica materials for laccase immobilization and tetracycline removal: A comprehensive study, *Microporous Mesoporous Mater.*, **291**, p. 109688.

Zhang, Y., Lam, F. L.-Y., Yan, Z. F., and Hu, X. J. (2009). Fabrication of copper (I) nitride nanorods within SBA-15 by metal organic chemical vapour deposition, *Sci. China Ser. E,* **52**, pp. 352–356.

Zhang, Y., Jiang, T., Zhang, Q., and Wang, S. (2010). Inclusion of telmisartan in mesocellular foam nanoparticles: drug loading and release property, *Eur. J. Pharm. Biopharm.*, **76**, pp. 17–23.

Zhang, Y., Zhang, J., Jiang, T., and Wang, S. (2011). Inclusion of the poorly water-soluble drug simvastatin in mesocellular foam nanoparticles: drug loading and release properties, *Int. J. Pharm.*, **410**, pp. 118–124.

Zhang, K., Yuan, E.-H., Xu, L.-L., Xue, Q.-S., Luo, C., Albela, B., and Bonneviot, L. (2012). Preparation of high-quality MCM-48 mesoporous silica and the mode of action of the template, *Eur. J. Inorg. Chem.,* **26**, pp. 4183–4189.

Zhao, C., Li, J., Chen, W., Yang, Y., Chiang, K., and Burke, N. (2015). Synthesis and electrochemical properties of ordered mesoporous carbon supported well-dispersed cobalt oxide nanoparticles for supercapacitor, *Mater. Res. Bull.,* **64**, pp. 55–60.

Zhu, K.-K., Yue, B., Xie, S.-H., Zhang, S.-Y., Zhang, B., Jin, S.-L., and He, H.-Y. (2004). Preparation and characterization of divanadium pentoxide nanowires inside SBA-15 channels, *Chin. J. Chem.*, **22**, pp. 33–37.

Zlateski, V., Keller, T. C., Pérez-Ramírez, J., and Grass, R. N. (2015). Immobilizing and de-immobilizing enzymes on mesoporous silica, *RSC Adv.,* **5**, pp. 87706–87712.

Chapter 4

Inverse Opal: Ordered Macroporous Inorganic Oxide Nanomaterials

Humphrey H. P. Yiu and Taylor Reid

School of Engineering and Physical Sciences, Heriot-Watt University, Edinburgh, EH14 4AS, UK

h.h.yiu@hw.ac.uk

4.1 Introduction

Liquid crystal templated nanoporous materials are fascinating, considering that the complexity of the materials can be made from these soft templates. However, their main limitation is within the constraints of the liquid crystal templates themselves. The dimensions of most liquid crystals are only several nanometers and any expansion normally leads to the loss of long-range order of the templated nanomaterial. Such attempts were demonstrated in around 2000 as a new class of materials called Mesostructured Cellular Foams, or MCFs, were discovered, as discussed in Chapter 3. However, MCFs do not have an ordered structure. The pore dimensions of these MCF materials are normally several nanometers, with a few exceptional cases up to 20 nm. In liquid crystal templating, ordered materials with pore

Nanotemplating: Engineering Nanocomposite Materials for the Future
Humphrey H. P. Yiu

ISBN 978-981-4316-92-7 (Hardcover), 978-1-003-50999-8 (eBook)
www.jennystanford.com

size beyond 20 nm are virtually unheard of. If an ordered nanoporous material is needed, we have to move away from liquid crystal templating routes.

Less than 10 years after MCM materials were discovered, scientists managed to use colloidal polymer nanoparticles to make nanomaterials with a 3-D nanoporous structure. And, of course, the pore dimensions of these nanomaterials can be at a 100 nm range and above, which is based on the size of the diameter of the polymer nanoparticle templates. The use of nanoparticles as a template for preparing complex nanoporous materials is commonly known as inverse opal. Such a method has a significant difference from methods introduced previously. Indeed, with this method, nanoporous materials are prepared with a "hard" template, while liquid crystals and surfactants are soft templates. Soft templates can be tricky to manipulate and self-assembly is the likely mechanism for the formation of nanoporous solids. With hard templates, we can deposit the template particles on a pre-fabricated platform for building nanomaterials with high complexity.

Most inverse opal materials were prepared from a nanoparticle template and a materials precursor, such as tetraethyl orthosilicate (TEOS) for silica, to form the base materials. Figure 4.1 illustrates the basic steps for inverse opal preparation. In this chapter, we will go through these steps in detail.

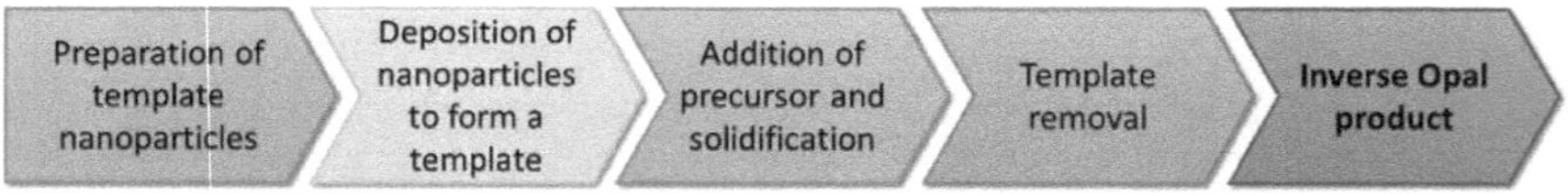

Figure 4.1 Key steps to the preparation of an inverse opal material.

There are several key points to be considered when we are going to prepare an inverse opal material. The template has to have a significant difference in chemical properties in order to enhance the removal of the template once the material is formed. Also, the surface chemistry also affects the framework formation of inverse opal materials.

4.2 Preparation of Colloidal Nanoparticles as Templates

In Chapter 1, we have already introduced several methods for preparing inorganic and organic nanoparticles using the surfactant or micelle-templating method. These nanoparticles are also ideal templates for preparing inverse opal materials because of their homogeneity and spherical morphology. However, as we have already pointed out, these nanoparticle templates are not always easy to make, can be prepared in only small quantities and are not always feasible to be scaled up. Since these templates are to be removed and discard eventually, we have to choose a nanoparticle template which is easy to make and to be removed.

In general, there are two main choices of template materials: organic polymer nanoparticles and inorganic crystalline nano-particles (see Table 4.1). Another option are metal nanoparticles, which are less common due to their high cost. The two key criteria for a suitable template are:

1. Ease and economic viability of production.
2. Ease of removal while causing as little damage as possible to the inverse opal nanomaterial structure.

Table 4.1 List of possible inverse opal template materials

Organic polymers	Inorganic oxide	Metals
Polystyrene (PS)	Silica	Platinum
Polymethyl Methacrylate (PMMA)	Titania	Palladium
	Zirconia	Gold
		Copper

The ideal shape of the template nanoparticles is spherical, which gives uniform pore size for both the inverse opal material and the residual voids once the template is removed. Spheres also can form a stable close-pack structure. Generally, the size of the template nanoparticles can be controlled by altering the reaction temperature and mixing speed, in addition to adding structure-directing reagents such as surfactants. The most common materials used for inverse opal templates are polystyrene and silica nanospheres, with polystyrene being the first template material ever used for this purpose.

4.2.1 Organic Polymer Nanoparticle Templates

The first inversed opal material (sometimes called 3DOM, or 3-D ordered macroporous) was prepared with an organic polymer template, polystyrene nanoparticles (Holland et al., 1998). The synthesis of polymer latex particles with several hundreds of nanometers in diameter has been known for a long time. Usually, an emulsion polymerization method is used for preparing latex particles of high quality and homogeneity. The size of these nanoparticles can be controlled simply by varying the reaction conditions such as speed of stirring and temperature of the reaction. For templating inverse opal materials, polystyrene (PS) and poly(methyl methacrylate) (PMMA) nanoparticles are the most popular candidates. This is because the narrow size distribution and the particle size of PS and PMMA nanoparticles can be easily tuned. Therefore, a wide variation of pore geometry and, more importantly, the bandgap can be optimized. These polymer nanoparticles can be removed either by solvation with a solvent (e.g., toluene or tetrahydrofuran) or by burning in air (calcination).

Box 4.1

Preparation of polystyrene latex nanoparticles with the micelle-templating procedure

Extreme caution has to be taken since many surfactants and monomers are toxic. This experiment must only be carried out by experienced synthetic chemists, with full risk assessment, COSHH and precautions in place.

There are many reported methods to prepare polystyrene latex particles that are suitable for use as the template for inverse opal. The method described here uses surfactant as a structure-directing agent. This is to give a sample with a narrow particle size distribution.

Chemicals:
Sodium dodecyl sulfonate (SDS, 0.18 g)
Purified styrene (15 mL)
Potassium persulfate ($K_2S_2O_8$, 0.1 g)

1. Dissolve all SDS in 50 mL of distilled water in a 100 mL two-necked flask with mechanical stirring under purging nitrogen.
2. Warm the SDS solution to 60°C under nitrogen.
3. Purified styrene was added to the surfactant solution dropwise, and the emulsion was stirred vigorously for 1.5 h until a stable emulsion was obtained.
4. Dissolve all $K_2S_2O_8$ in 1 mL of water.
5. Add the $K_2S_2O_8$ solution slowly to the emulsion with stirring.
6. The reaction temperature was raised to 80°C and the emulsion was allowed to react for 24 h under nitrogen.
7. A stable white emulsion of PS particles can then be obtained at the end of the reaction.

4.2.2 Template-Free Synthesis of Silica Nanoparticles (the Stöber Method)

In addition to polymer nanoparticles, inverse opal materials can also be templated using inorganic nanoparticles for templating. Silica nanoparticles are among the most popular candidates for this application. Silica is a stable, amorphous inorganic solid (with the exception of a few crystalline phases such as quartz). Therefore, silica can easily be "molded" into designed shapes. Also, silica nanospheres can be grown relatively easily compared with other inorganic oxides. Another merit of silica is its acidic nature, which is in contrast to most metal oxides. Therefore, the removal of silica using a strong base does not have a devastating effect on the prepared nanoporous metal oxide material.

The preparation of silica nanoparticles has been known for decades. The Stöber method, developed by Stöber in the 1960s, is a template-free method. The experimental procedure is simple and easy to scale up. Also, apart from ammonia, there is no dangerous chemical involved in the synthesis. The Stöber method is widely used to prepare silica nanoparticles for various uses.

Box 4.2

Preparation of silica nanoparticles using the Stöber method

Extreme caution has to be taken since ammonia can be corrosive and harmful if inhaled. This experiment **must be carried out by experienced synthetic inorganic chemists only**, with full risk assessment, COSHH and precautions in place.

There are many variations of the Stöber method since it was first published in 1968. Here is just one of the examples:

Chemicals:

Tetraethylorthosilicate TEOS: 5.6 (mL)
Ammonia NH_4OH(29%): 9.8 (mL)
Deionized water: 10.8 mL
Ethanol EtOH: 73.8 mL

Procedure:

1. Mix the ammonia, DI water and ethanol to form a homogeneous solution with a magnetic stirrer.
2. Add TEOS slowly under stirring.
3. React for 12 h.
4. Collect the colloidal particles by centrifugation.
5. Wash particles by re-disperse in pure EtOH with sonication and recovered by centrifugation, repeat several times (> 3).
6. Wash with water following the same procedure as step 5 until pH reaches 7.
7. Characterize samples with SEM.

4.3 Deposition of Template Particles into 3-D Patterns

The deposition process determines the arrangement of the template particles and hence influences the microstructure of the final inverse opal materials. Therefore, choosing the right deposition process can greatly affect the quality and properties of the prepared materials.

4.3.1 Gravity

The easiest way to arrange the template particles is to exploit the natural gravitational force. The use of gravity allows the spherical nanoparticles to arrange themselves into a closely packed pattern naturally, with a 3-D hexagonal structure. When a small volume of nanoparticle suspension is allowed to stand in a mold under gravity, the nanoparticles will settle down to the bottom of the mold slowly. However, this is a slow process, in particular for template materials of low density, such as organic polymers. Evaporation of the liquid phase, usually water, can speed up the deposition process. Even that, deposition using gravity can take days to complete.

4.3.2 Centrifugation

Making use of centrifugation can speed up the deposition process. Since the G-force produced by centrifugation is a few times higher than that of gravity, the deposition process can be completed in 10 min. However, we cannot have a lot of choices in the shape of molds. The deposition process has to be carried out in centrifuge tubes and mostly they are simple round-bottomed, cylindrical tubes. Other shapes such as flat films of inverse opal materials are almost impossible. The use of centrifugation does give us a good testing ground for the fast preparation of new inverse opal materials. Also, the template materials need to have a higher density than the solvent. Otherwise, the template cannot assemble.

In general, a suspension of template particles, e.g., polymer latex particles, is added to a number of centrifuge tubes and placed inside a centrifuge. The number of tubes, the shape and size of tubes depend on the centrifuge machine. Larger centrifuge machines can take as many as 16 tubes of 50 mL of suspension. Such high volumes are unlikely to be achieved using gravity and ambient drying in a short time. The speed of centrifugation may vary from 2000 to 13000 rpm. A faster speed is likely to work better for lighter template materials and the template forms faster. However, this may also lead to a less ordered structure.

Once the particles are separated out and the template is formed, the supernatant can be discarded. If a larger template is needed, a second round of deposition is possible and so on.

4.3.3 Suction-Aided Deposition

In order to speed up the deposition process, inverse opal templates can also be prepared under suction. This method allows the preparation of a flat template and works well for light template materials. Also, unlike centrifugation, the liquid phase is removed during the suction process, leaving behind the solid phase template particles. Ideally, a smooth membrane filter is needed but this may slow down the process. There are many sizes of membrane filters available to choose from. Hence, suction is a popular method for preparing inverse opal templates. However, that is also a limitation of suction. Notably, the membrane filters are also the only option for template preparation. As a result, unlike preparation through gravity, only flat templates can be prepared by suction.

4.3.4 Self-Assembly

Like many species in nature, molecules can self-assemble through interaction with each other. Such a phenomenon can also be simulated using surface-modified nanoparticles. This can be achieved with electrostatic interaction, chemical binding and biomolecular interactions. Self-assembly can create an inverse opal template with a higher degree of complexity. For example, a bimodular template using nanoparticles of two different sizes, one larger and positively charged, the other smaller but negatively charged, can be assembled. The materials cast at the end will possess a bimodular porous system. Inverse opal materials with such a complex structure can only be achieved by using designed assembling mechanisms.

In addition to electrostatic interaction, stronger interaction can be used for self-assembly. Chemical binding is a popular choice. For example, amine- and aldehyde-functionalized surfaces can bind with each other to form imine bonds in ambient conditions without further modification. However, in recent years, we have seen more and more effort in exploiting biomolecular interactions. This can be termed a bio-inspired assembly. One common example is the use of single-strand DNA to decorate nanoparticles. DNA molecules are polymeric nucleic acids with only four base units, A, T, C, and G. Since A binds strongly with T and C does the same with G, carefully chosen sequences couple with each other. It was found that a sequence of 12 bases (e.g., A_{12} vs T_{12}) will provide strong interaction for stable assembly. The main downside of using assembly with biomolecular interaction is that these bio-species are very expensive when compared with most chemical species. Therefore, it is only likely to be used in preparing products of an extremely high value. Other biomolecular interactions that can be exploited include antibody-antigen interaction and the biotin-streptavidin pairing.

4.3.5 Patterning and Pre-Fabrication

In order to manufacture materials to suit specific applications, micro-fabrication is essential by building components on a miniature scale. The use of "stamps" has been common in preparing materials for nanoelectronic and nanobiotechnology. However, nanostructures within the fabricated materials will increase the complexity of the final product. This can be achieved by combining the inverse opal templating with micromolding using stamp fabrications. Advances in 3-D printing in recent years also help to design pre-fabricated molds with specific structures that suit the application. With these designed molds, nanoparticle templates can be deposited, preferably with gravity. As the resolution of 3-D printing gets higher, the complexity of pre-fabricated molds can also increase.

4.3.6 Magnetic Patterning

The exploitation of a magnetic field in preparing inverse opal templates is still a new concept. Logically the template particles have to be magnetically active, such as iron oxides. In recent years, significant advancement has been shown in preparing homogeneous magnetic nanoparticles in large quantities (up to grams). This helps us to develop new magnetic methods for template preparation.

By placing a strong magnet underneath a suspension of magnetic nanoparticles, an ordered template will be assembled at a faster rate than using gravity solely. These magnetic templates can be used as those prepared with other methods. A unique advantage of using the magnetic field for making templates is that a template can be made against gravity. For example, the template can be made on the side of a container by placing the magnet at the chosen position. This feature can enhance the drying process after the templates are assembled. But more importantly, during the loading of precursor for inverse opal, a more controlled loading can be achieved by immobilizing the template above the bottom of the container and hence excess precursor can be easier removed by gravity. Further details of this feature will be discussed in the next section.

4.4 Choosing and Deposition of Precursors

Apart from the templates, choosing the precursors also needs to be carefully considered. Of course, such choice is highly dictated by the final products of interest. For example, if the final inverse opal is to be a TiO_2-based material, a titanium precursor must be used. Also, we should consider how to transform a liquid phase/soluble precursor into the solid phase we aim for. Therefore physical and chemical properties of a precursor need to be carefully examined and see if it is compatible with the template material.

Moreover, the surface chemistry of the template material can also affect the formation of an inverse opal material, and therefore the choice of precursors. One good example is the use of polystyrene (PS) latex particles. The hydrophobic surface, due to the benzene ring on the styrene monomer units, discourages the even distribution of precursors unless an oil-based precursor is used. However, most of the precursors for inorganic oxides are hydrophilic and water soluble. The formation of a thin layer of precursor is an ideal way to begin the deposition of precursors. To overcome this compatibility problem, surface modification of the template can improve the formation of the first thin layer of precursors. For example, PS modified with carboxylic acid groups enhanced the interaction between the precursors and the surface of template particles. The carboxylic acid groups not only improve the hydrophilicity of the nanoparticles but also catalyze the hydrolysis and condensation of precursors. Another example is the use of thiol-functionalized silica particles for strong interaction with metal nanoparticles and further metal deposition.

The deposition of precursors to the inverse opal templates needs to be simple. Therefore, it is a good idea to choose a precursor which can decompose/solidify under mild conditions (room temperature, 1 atmosphere). Usually, ethanol solutions of alkoxides are preferred choices because they can be "dried" in 24 h. After drying, heating with a slow heating rate (e.g., 2°C/min) is usually carried out to ensure the complete decomposition of the precursor. For preparing silica inverse opal, TEOS is the most popular choice because it is widely available and widely used for preparing other silica nanostructures including SBA-15 and for Stöber synthesis. Titanium alkoxides, such as titanium(IV) isopropoxide, can be used to prepare titania TiO_2 inverse opal. However, the use of alkoxides is limited to only a few materials because of their availability and cost. For example, iron(III) ethoxide is currently available but at $300 per 10 g, nearly ×400-500 times the cost of TEOS, making them economically unviable.

Use of alkoxides as the precursors will need to undergo steps of "solidification," including evaporation of the solvent, drying and then crosslinking of precursors upon decomposition. As suggested earlier, ideally, these steps should be carried out under mild conditions. Figure 4.2 illustrates the chemistry of the decomposition of an alkoxide and its crosslinking. The final heating step can ensure complete decomposition. For polymer-based inverse opals using SiO_2 nanoparticles as a template, a precursor mixture of monomers and initiators is added to the template and the polymerization can be initiated with UV radiation.

Another issue needed to be considered is that the structure/assembly of templates are needed to be kept intact during precursor deposition. Therefore, deposition might be helped by suction. This can be less critical if the assembly of templates was formed by electrostatic self-assembly and, even less so, chemical binding, since the template structures from these methods would be more robust.

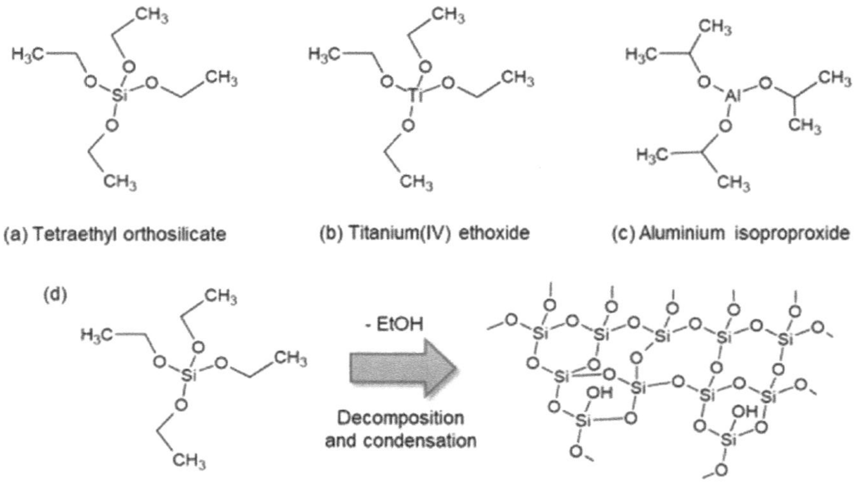

Figure 4.2 Structures of common precursors for inverse opal (a) tetraethyl orthosilicate (TEOS), (b) titanium (IV) ethoxide, (c) aluminum isopropoxide. (d) an illustration of the decomposition reaction of TEOS to form silica.

4.5 Template Removal

In order to "reveal" the porous structure of the inverse opals, as seen in other templated nanoporous materials, the template must be removed after the solidification of precursors. Similar to the removal of surfactant templates for organic mesoporous silica syntheses, there are two standard routes for organic template removal in inverse opal synthesis: (i) calcination and (ii) solvent extraction. Of course, the method chosen has to keep the inorganic inverse opal structure intact after template removal. Generally, solvent extraction is a "milder" method and the risk of structural collapse is lower than calcination. For example, for PS templates, tetrahydrofuran (THF) can be used as the solvent and extraction would usually be carried out near the boiling point of the solvent to maximize the speed of extraction. However, it is usually less efficient and complete removal (extraction) of the template can be difficult because the solvation rate of polymer templates (PS or PMAA) can still be low. Therefore, the extraction method tends to be more time consuming and labor intensive. Complete removal of the template using solvent extraction may need repeat extraction several times. Such a procedure can take up to a week.

In contrast, calcination is more aggressive and likely to achieve complete removal of the template, leaving desired cavities in the final inverse opal structure. At a high temperature of calcination (say 700–800°C), inorganic oxide can contract and therefore the pore size may reduce. Calcinations usually cause the shrinking of materials and the pore dimension is likely to be reduced up to 10%. Moreover, structural damage can be observed due to the high temperature involved in calcinations. As a result, calcination is best avoided for the preparation of delicate inverse opal materials or large monolithic solids. Nonetheless, calcination is still the preferred route for the removal of polymeric templates.

Among the polymer templates, there are some subtle differences in the way they are being decomposed and removed during calcinations. For PS templates, the polymer burns like many polymeric materials do. PS combusts at high temperatures and gives out high energy. Structural damage to materials is caused by the sudden release of gas and high energy. In contrast to PS, PMMA templates start to decompose at a lower temperature (typically around 250°C) by slowly breaking the polymer into small organic monomer units. These small units escape from the material as a gas and the residual organic matters are to be burnt at a higher temperature. This is a much milder process than high-temperature combustion. As a result, scientists prefer to use PMMA as a template for inverse opal materials once they realized the benefit of low-temperature calcination.

Calcination also serves as a condensation process for converting inorganic precursors into the desired inorganic oxides. For example, metal precursors such as acetates and nitrates need to be heated up to around 250°C to form the corresponding metal oxides. However, other precursors such as alkoxides (e.g., tetraethylorthosilicate or TEOS, titanium tetrabutoxide and aluminum tripropoxide) decompose in the air and may not need any heating at all. Unfortunately, not many metals form stable alkoxides to be used as a precursor. Therefore, the use of acetates or nitrates is the common way of preparation. Other salts such as chlorides and sulfates are less common to use due to possible residues remaining in the structure after calcination.

Removal of templates with calcination is only applicable to organic templates and inorganic carbon templates. Most inorganic templates (e.g., SiO_2) do not combust and these require extraction or dissolution for removal. This makes the choice of template materials complicated as many inorganic materials have similar properties including being soluble in strong acids and insoluble in organic solvents. As a result, some combinations of template and "wall" materials can be difficult to separate. For example, if an inverse opal titania material was prepared with iron oxide nanoparticles, the template cannot be removed with organic solvents (neither dissolves) or strong acids (both dissolve). This highlighted the importance of the choice of template materials.

Apart from organic polymer templates, silica nanoparticles are also popular, thanks to their acidic nature, unlike most metal oxides. Therefore, silica templates can be removed by using strong alkalis (e.g., 5 M NaOH, KOH or even sodium ethanoxide, which is stronger than its hydroxide cousin). Nonetheless, silicas are notorious to be dissolved and total removal or silica templates using NaOH, for example, is not a guarantee. Hence, many reported methods would prefer more aggressive reagents such as hydrofluoric acid (HF). HF is an extremely dangerous chemical and can cause serious damage to human bodies (HF is capable of dissolving bones!). Maximum precaution measures are required when HF is used. Other methods for template removal include using photodegradation and biodegradation of polymers, and bio-etching of inorganic materials.

4.6 Characterization Methods for Inverse Opal

4.6.1 Scanning Electron Microscopy

The technique of scanning electron microscopy (SEM) has already been introduced in Chapter 1. For many nanostructured materials with a <100 nm scale, e.g., mesoporous silicas, SEM is only useful to study the particle morphology (outer structure). Its resolution is still not high enough to examine the long-range order at the nanoscale. However, for inverse opals where the "ordered structure" are at a scale of ca. 100 nm, SEM becomes a powerful tool to examine their structure. The uses of SEM for inverse opal can be in two stages.

First, when the nanoparticle templates were made, SEM is used to check their size and homogeneity, ideally after the template is cast. This would provide some insight into the structure of the final product before the addition of the precursors. The second, and more important, use of SEM for characterizing inverse opals is examining the porous structure of the final products. A distinctive pattern will be shown due to the packing of the nanoparticle templates, as shown in Fig. 4.3. For most inverse opal studies, a standard modern SEM that can allow a magnification of around ×10,000 would be suitable. If the nanostructure in the wall of the inverse opal materials is of specific interest, such as in trimodal inverse opals that will be discussed in Section 4.8, a transmission electron microscope (TEM) will then be required for a high enough magnification.

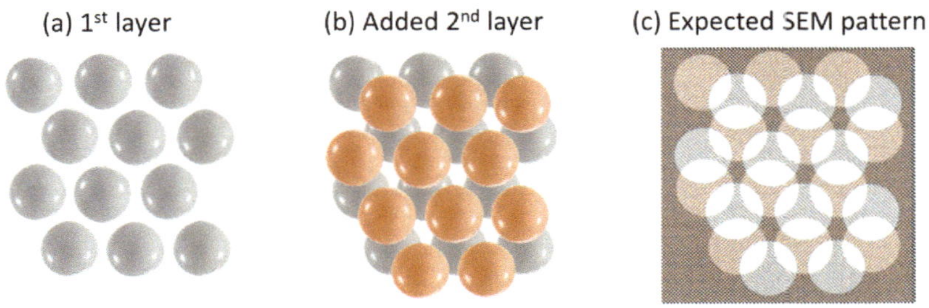

Figure 4.3 Illustration of how a distinctive pattern of inverse opal was shown in SEM.

4.6.2 UV-Vis Spectrometry and Bandgap Measurement

One important parameter of inverse opal samples that needed to be measured is their band gap. This relates to the absorption wavelength of the materials, lower band gap energy corresponding to a longer wavelength. The band gap of photonic materials, including inverse opals, can be measured using a UV-vis spectrometer for suitable solid-state samples by recording their diffuse reflectance spectra. Usually, a UV-vis spectrometer capable of measuring solid-state samples is a lot more expensive but these units should have built-in functions to convert a diffuse reflectance spectrum (R% vs wavelength plot) into a *Tauc plot*. For a semiconductor or inverse opal sample, the Tauc plot (see Fig. 4.4) is needed to determine the band gap. We also need to

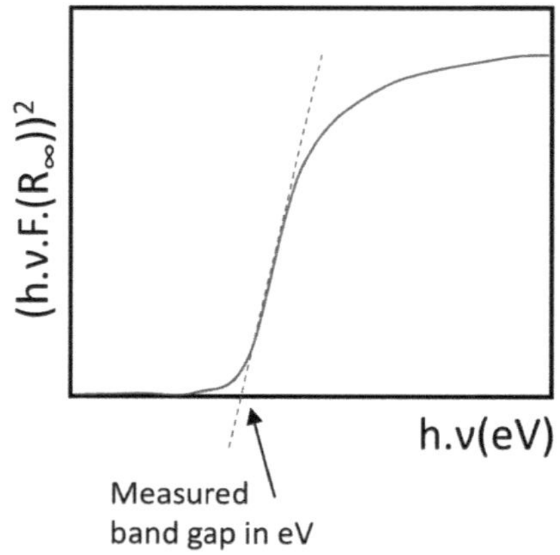

Figure 4.4 A typical Tauc plot for a semiconductor sample for the determination of band gap.

Table 4.2 List of band gap values for several common oxides

Materials	The band gap (eV)
TiO_2	3.2
WO_3	2.7
$BaTiO_3$	3.3
SnO_2	3.5
Bi_2O_3	2.9

be careful that the band gap of a material can easily be wrongly determined due to the complexity of a Tauc plot (Makuła et al., 2018). Table 4.2 summarizes the band gap values for a few common oxide materials.

4.7 Applications of Inverse Opal Materials

4.7.1 Electrical and Electronic Applications

Conductive metal materials with a high surface area would have a wide range of electrical and electronic applications. The synthesis of metallic (copper and nickel) inverse opals has been reported for this purpose using sulfate-terminated polystyrene spheres as the templates (Barako et al., 2016). Interestingly, the porous structure had a different effect on the thermal conductivity of the inverse opal; while the pore diameter (from 100–1000 nm) of the copper inverse opals enhanced the conductivity (up to 60 $Wm^{-1}K^{-1}$), the conductivity of nickel inverse opals saturated at 15 $Wm^{-1}K^{-1}$ across the whole range of pore size. Such observation suggested that the relation between the physical properties and the pore size of an inverse opal is rather complicated.

Semiconducting inverse opals are also highly sought after but they are not always easy to make. For example, silicon inverse opal has been reported by reducing silica inverse opal. However, a very strong reducing agent is required to transform SiO_2 to Si. For example, Hatton et al. demonstrated using Mg vapor as the reducing agent (Hatton et al., 2010). Another pathway to synthesize Si inverse opal was using chemical vapor deposition (CVD) onto a template with silane (Si_2H_6) as the silicon source (Blanco et al., 2000). Although this is technically difficult, other semiconducting inverse opal materials such as germanium (Ge), as well as layered inverse opal can also be made from this method (Tétreault et al., 2004).

In addition to metals and semiconductors, many metal nitrides also have important electrical and electronic applications since they have metallic properties at visible wavelengths. In particular, transition metal nitrides such as titanium nitride and zirconium nitride are also stable, shiny in appearance and hard. Remarkably, their optical properties can be tuned by changing their compositions. Their applications in electronics include use as gate materials to *n*-type and *p*-type transistors. These nitrides also exhibit good superconductivities at low temperatures (<10 K, or −263°C). Therefore, nanostructured metal nitrides would be very useful in electronic industries. Using inverse opal for templating metal nitride is aimed to enhance the specific surface area of the material, a similar outcome to that of liquid crystal templated metal nitride as we learned in Chapter 3. However, inverse opal is mechanically stronger than liquid crystal templates. In Chapter 3, we also learned that these metal nitrides can also be made using second or third generation via hard templates from the liquid crystal route. Inverse opal (polymer) templates can be removed by solvent extraction, which is a milder template removal regime and less damaging to the final metal nitride product. Nonetheless, the polymer latex particles of the inverse opal template can also be removed by calcination (or burning), but that will cause similar damage as combusting the carbon CMK hard template for preparing the 3rd generation of porous metal nitrides.

Nanostructured TiN inverse opal was reported using polystyrene (PS) particles and titanium(IV) dimethylamide as the template and titanium precursor (Gray et al., 2009). The macroporous inverse opal structure was formed by direct nitridation of the organometallic precursors at 600°C under a flow of dry ammonia. The PS template was removed at this high temperature with no carbon residue. In order to fine-tune the properties of nitrides, scientists tend to alter the composition by doping a different element to the structure. For example, doping tungsten to form Ti_xW_yN has been a successful example with enhanced conductivity. It is also possible to prepare nitrides by nitridation of the corresponding metal oxide. For instance, $Ti_{0.7}W_{0.3}O$ were successfully synthesized in the inverse opal morphology using the carboxylate-functionalized PS beads ranging from 50, 100, 200 and 500 nm (Subban et al., 2012). The $Ti_{0.7}W_{0.3}N$ inverse opal was formed following the nitridation of the inverse opal oxides and retained the long-range microporous morphology in a single crystal phase. Interestingly, this nitride inverse opal can be re-oxidized back to the oxide with a cycling reaction.

4.7.2 Photocatalytic Applications

As we have seen in previous chapters, catalysis is one key area to be considered for the development of new materials. Inverse opals have also been considered for catalytic application early on. However, siliceous inverse opals (SiO_2) have very little catalytic activity without surface modification or doping. Furthermore, using inverse opals as simple catalysts carriers would not show significant improvement in catalytic activities in comparison with other large pore supports (e.g., MCF) of the same material since the pore size in inverse opal is usually much larger than most reactant and product molecules. Mass transfer in channels becomes a less significant issue. Therefore, scientists turned to other materials that are intrinsically catalytic.

The superb photonic properties of inverse opals are a good match for photo-related applications, notably photocatalysis. Titania (TiO_2) is considerably the most popular photocatalyst that has been widely studied in research and used in industries. The two phases (anatase and rutile) are both catalytic active, but anatase showed a high activity usually due to a larger bandgap (Guo et al., 2019). They are widely used in the degradation of organic pollutants, dyes, and toxic chemicals. With the enhanced optical property, an inverse opal structure of TiO_2 has been shown to improve the catalytic activity. To assess the photocatalytic activity of a TiO_2 sample, degradation of a dye is commonly performed due to the ease of measurement using UV-vis spectrometers, as well as the visible effects for demonstration. For example, in a study on the degradation of phenol, inverse opal TiO_2 catalyst showed an improved activity over the disordered macroporous TiO_2 and an inverse opal sample that has its 3D structure destroyed mechanically, confirming the effect of the 3D inverse opal structure on photocatalytic activity (Sordello et al., 2011). It was also found that the bandgap of the inverse opal structure (e.g., the size of the cavity) can further influence the activity (Chen et al., 2006). In a separate study on the degradation of Acid Orange 7 (an azo dye), doping the TiO_2 inverse opal with Ti^{3+} ions can enhance the photocatalytic activity even under visible light (Qi et al., 2014). This is significant because conventional TiO_2 photocatalysts can only utilize around 5% of the energy of sunlight. Shifting the absorption towards the visible spectrum can improve the energy efficiency of the photocatalytic processes.

Another method to improve the light absorption of TiO_2 towards the visible spectrum is to support noble metal particles on the surface, notably platinum (Pt). For example, Liu et al. demonstrated using Pt-TiO_2 inverse opal catalysts for a "water splitting" reaction to generate hydrogen (Jian et al., 2010). Hydrogen is currently considered to be a valuable fuel because it does not generate greenhouse gas, particularly CO_2, when it is combusted to generate energy. Conventional hydrogen production from steam reforming is highly energetic and also generates a large amount of CO_2 as by-products. A low-energy hydrogen production method with zero (direct) carbon emission will be welcomed by industries and governments globally. In this study, the inverse opal structure enhanced hydrogen production by about 2.5-fold when compared with a standard nanocrystalline Pt-TiO_2 catalyst. Inverse opals of other oxides such as WO_3 and SnO_2 have also been reported. However, at around 2 mmol after 8 h using 10 mg of catalyst, the hydrogen production rate is still very low to be useful on an industrial scale. In another study related to fuel and the environment, Pt-TiO_2 inverse opal can transform the greenhouse gas CO_2 to methane (CH_4) in water under visible light (Jiao et al., 2017). The conversion rate was very slow, at 2.4 μmol in 8 h with 0.1 g of catalyst. More importantly, the product was predominately methane, which is a worse greenhouse gas (by an estimated ×2.5 times) than CO_2 itself.

4.7.3 Biomolecular Detections

The photonic properties of inverse opals have been exploited for the detection of biomolecules with high specifications. The principle is based on the blue/red shifts on the photonic bandgap due to the bound analyte in the binding cavities by changing the refractive index. In order to achieve specificity in detection, antibodies may use for binding the targeted analytes. Figure 4.5 illustrates this basic principle.

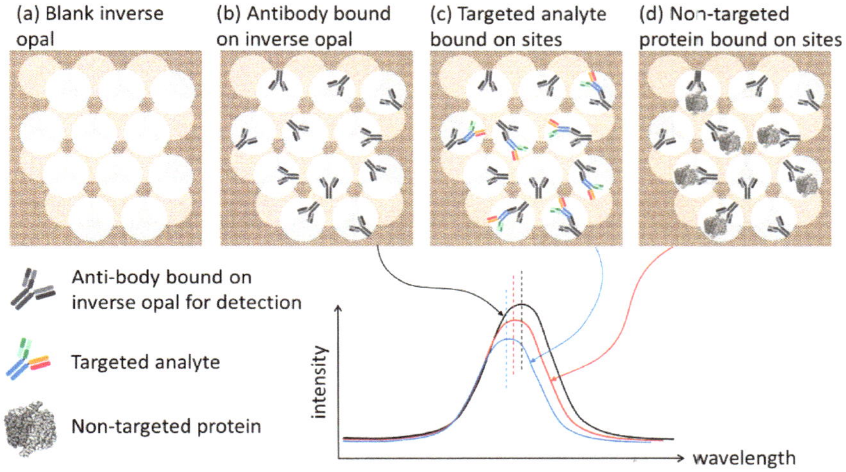

Figure 4.5 Illustration demonstrating the principle of biomolecular detection using inverse opal.

The use of inverse opal for biomolecular detection offers several distinctive advantages:

1. High surface area of inverse opals allows high binding capacity.
2. Highly ordered structures give superior optical properties.
3. Simple functionalization methods for grafting sensing molecules such as antibodies.
4. Optical properties (e.g., emission wavelengths) can be tuned by tuning the pore size.

Point 4 here is particularly interesting because the pore size of an inverse opal can have a very good correlation between pore size and emission wavelength (see Fig. 4.6) (Shen et al., 2012). Since there are many methods to tune the particle size of the polymer templates, inverse opals of different wavelengths can easily be made. As such, multi-target detection can be achieved using inverse opals.

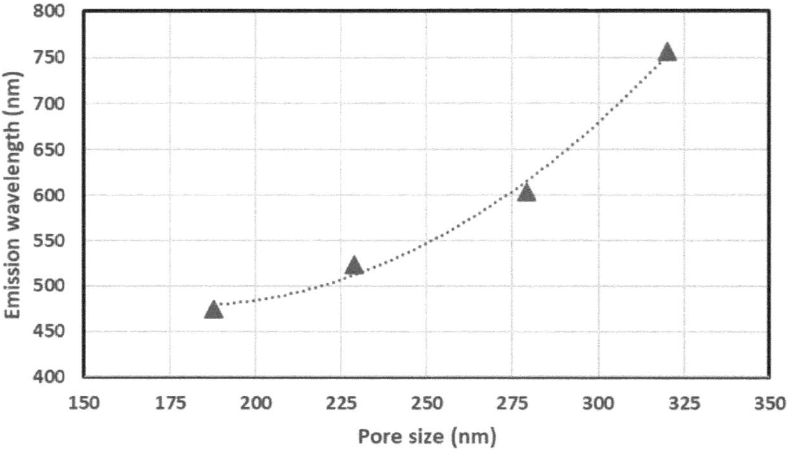

Figure 4.6 Correlation between the pore size and the emission wavelength of a TiO_2 inverse opal.

Detection of proteins or peptides is always of great interest due to their importance in biomedical diagnosis. Moreover, their wide variety but similarity in structure possesses significant challenges to scientists to develop accurate monitoring with fast response time. Inverse opal can offer a solution for this purpose. Further challenges can be due to the complexity of samples, e.g., blood and other bodily fluids, where there could be thousands of different biomolecules and the target molecules can be of a rather low concentration. Li et al. demonstrated the detection of human immunoglobulin G (IgG) using TiO_2 inverse opal decorated with anti-human IgG antibody inside the pores (Li et al., 2008). Similar to the principle illustrated in Fig. 4.5, a blue shift was observed and measured when human IgG was bound to binding sites. The sensitivity was shown to be as high as 1 μg/mL. Liver cancer marker alpha-fetoprotein (AFP) was also shown to be detectable using inverse opal constructs, with a remarkable detection limit down to 10 pg/mL (Zhao et al., 2015).

In biomedical diagnoses, other biomolecules of interest include DNA and carbohydrates. For the detection of a specific DNA sequence, the detection principle using inverse opal can be very similar to that demonstrated for proteins. For example, Chiappini et al. have demonstrated the detection of a single-strand DNA (ssDNA) with a specific sequence using SiO_2 inverse opal functionalized with DNA-aptamer and Cy3 fluorophore (Chiappini et al., 2018). Another example used hydrogel inverse opal functionalized with ssDNA and quantum dots (QDs) (Zhao et al., 2010). This system showed a detection limit of 1 nM. The detection limit can be improved by 100-fold using gold nanoparticles with localized surface plasmon resonance (LSPR) technique instead of QDs (Shen et al., 2012).

One key diagnostic application that inverse opals can have a huge potential is the detection of the whole virus. In recent decades, viral diseases, such as AIDS, SARS, MERS, Ebola and more recently Covid-19, have been major concerns due to the high fatality and contagiousness. The size of viruses ranges from 20 nm to 400 nm, which is perfectly compatible with the pore size of inverse opal. For instance, SiO_2 inverse opal decorated with antibody was demonstrated for detecting influenza A virus H1N1, which triggered a red shift upon binding (Lee et al., 2018). A high sensitivity (10^3–10^5 PFU) and selectivity were shown towards the target virus. Another example demonstrated using inverse opal hydrogel (templated using colloidal SiO_2 particles) functionalized with antibodies for the detection of Rotavirus (Maeng et al., 2016).

Detection of smaller biomolecules for healthcare applications is also possible using inverse opals. Since diabetes becomes a major concern in many Western societies, the detection of cholesterols using inverse opal sensors has been of interest. For example, carbon inverse opal, templated using SiO_2 nanoparticles, has been fabricated for the detection of cholesterols in blood samples (Zhong et al., 2015). A detection limit of 1.3–2.6 mg/mL was shown. There are also examples for diagnosis based on enzymatic catalysis using inverse opals and they will be discussed in the next section.

4.7.4 Other Biomolecular Applications

As seen in Chapters 2 and 3, the increase in the pore size of a material will allow the development of carriers for larger functional biomolecules. Enzymes are among the most popular molecules to be benefited from such development. We can see that use of mesoporous silicas for enzyme immobilization is very limited due to their relatively small pores. The expanded pore size of MCF enhances their use for this purpose. Naturally, inverse opals with pore sizes between 100–400 nm would be studied for enzyme immobilization. However, many of the examples of using inverse opals for enzyme immobilization in the literature were for sensing purposes.

Glucose oxidase GOx can oxidize D-glucose using O_2 (from air) to gluconolactone and H_2O_2 is released as a by-product. The released H_2O_2 is widely used for sensing/quantifying D-glucose. GOx is a rather large molecule (ca. 160 kDa); many mesoporous materials are not suitable as their carriers. The large pores of inverse opal materials become useful for GOx immobilization. You et al. demonstrated immobilization of GOx on a ZnO inverse opal for photoelectrochemical detection of glucose (You et al., 2013). It has shown a large linear detection range of glucose (0.01–18 mM), much wider than using other ZnO nanostructures.

Wang et al. demonstrated enzyme immobilization using an inverse opal hydrogel embedded with magnetic nanoparticles (Wang et al., 2017). Three enzymes (HRP, amylase and urease) were tested for immobilization, showing the versatility of this method. Instead of conventional immobilization, Jiang et al. incorporated enzymes (HRP and amylase) in bovine serum albumin (BSA)-based inverse opal (Jiang et al., 2014). The enzymes were crosslinked with BSA molecules to form a rigid inverse opal porous structure. This research group also demonstrated the immobilization of Penicillin G acylase (PGA) on a BSA inverse opal via a glutardialdehyde linkage after the polystyrene template was removed.

Similar to other porous materials we have introduced in previous chapters, the use of inverse opals for medical treatments, such as controlled drug delivery, has been studied in the literature. However, due to their rigidity, inorganic oxide inverse opals are less suitable for controlled delivery. More flexible polymeric inverse opals are therefore preferred. For example, an inverse opal material cast with poly(N-isopropyl acrylamide), or poly-NIPAM, has shown thermal response and release of drugs due to material contraction (Zhang et al., 2015). Upon drug release, a shift was observed in the reflection wavelength from the inverse opal. This phenomenon is similar to that used in biomolecular sensing and also allows for monitoring the drug release process. Another mode of stimuli being exploited was electrical release, demonstrated using an inverse opal cast with a conductive polymer, polypyrrole, loaded with an anti-inflammatory drug (dexamethasone phosphate) (Seyfoddin et al., 2015). This inverse opal was found to have a higher drug loading capacity and greater electrical response due to a higher surface area, when compared with the non-porous counterpart. One key criterium for use as carriers for drug delivery *in vivo* is that the carrier material (e.g., the inverse opal in this case) has to be non-toxic and biocompatible. However, such materials that are safe and suitable for casting inverse opals are not easily found. Therefore, examples of successful drug delivery using inverse opals are still limited in the literature.

4.8 Inverse Opal of Higher Complexity and Trimodal Porous Materials

Scientists always try to develop new materials with architectures of higher complexity and wish they could offer superior properties over competitors. Inverse opals are no exception. Very soon after the first report of inverse opal, a group of scientists in Germany had shown that if there are two polymer nanoparticle templates of very different sizes, the smaller nanoparticles can fill inside the voids generated in between the larger nanoparticles, creating a more complex template for forming inverse opal. Wang et al. demonstrated to use large PS nanoparticles template (465 nm) mixed with a smaller PMMA nanoparticle template (84 nm) to prepare a siliceous inverse opal material, as illustrated in Fig. 4.7) (Wang et al., 2006). The resultant materials showed a "binary" porous structure, created by the two different template particles. Although there was no application demonstrated, the study did pave the way for further development in multimodal inverse opal synthesis.

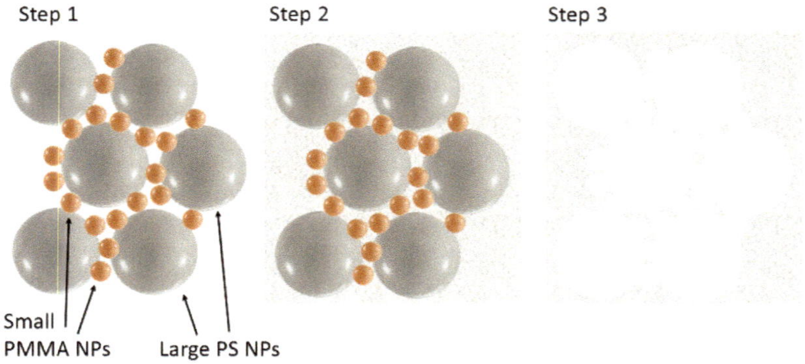

Figure 4.7 Inverse opal prepared by two templates of nanoparticles of different sizes.

A trimodal material was reported by taking advantage of the fact that SBA-15 type of materials had already shown a bimodal porous structure (microporous-mesoporous) and reported an inverse opal material with a "trimodal" porous structure (Costacurta et al., 2007). The preparation of this material is essentially the same as standard inverse opal materials but the precursor is a mixture of oxide precursor and Pluronic surfactant, as shown in Fig. 4.8. The resultant trimodal silica material did show a high surface area, ca. 550 m^2g^{-1}, a significant enhancement to those standard inverse opal silicas. There are pros and cons of such sophisticated silica material. Although it has a highly elegant structure, it does lose the "pore specificity" of ordered mesoporous silica such as SBA-15. The increase in complexity for synthesis may also outweigh any enhancement from the improved structure. Finally, siliceous materials are rather inert chemically and further functionalization is still required, further complicating the synthesis procedure.

In contrast to silica, titania (TiO_2) is intrinsically active as shown in many photocatalytic applications (as seen in Section 4.4.2). This method can be adapted for preparing a trimodal TiO_2, which would be a more useful material without further functionalization. For example, Mandlmeier et al. demonstrated the preparation of a trimodal TiO_2 material for possible photocatalysis (Mandlmeier et al., 2011). Photocatalytic degradation of rhodamine B using trimodal TiO_2 inverse opal was demonstrated a few years later by Liu et al. (Liu et al., 2015). The rate constant for degradation has been increased by ca. 6-fold compared with TiO_2 inverse opal with a non-porous wall. In this work, PEG (m.w. = 2000) was used as a structural directing reagent instead of a Pluronic surfactant. Nonetheless, the principle is demonstrated.

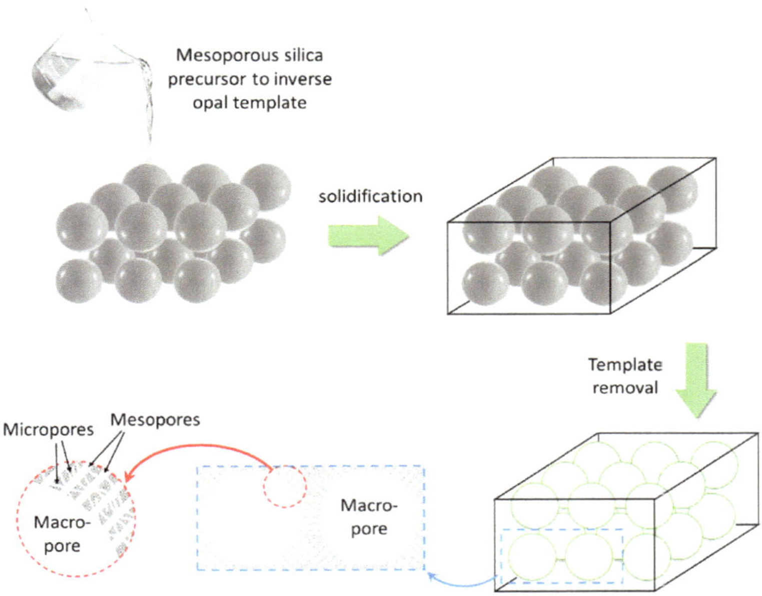

Figure 4.8 An illustration for the preparation of trimodal macro-meso-micro-porous silica based on the inverse opal template principle.

4.9 Biodegradable Inverse Opals and Templates

So far, since Chapter 2, most of the materials that we introduced are hard inorganic materials with ordered porous structures. This is partly because, without the template, soft polymer materials may not be mechanically strong enough to retain the delicate porous structure. However, it would be useful to have ordered porous materials made from organic polymers to widen the application range of ordered porous materials, including inverse opals. Very early on, a Japanese research group reported a poly(lactic acid) (PLA, or PLLA)-based inverse opal material (Musashi et al., 2007). The aim was to develop a highly porous implant material which was also biodegradable. PLA is known for biomedical applications while the porous structure of an inverse opal enhances the diffusion of nutrients for cell growth. More remarkably, the optical property of this PLA inverse opal showed strong near-infrared (NIR) reflection at ca. 740 nm. Such a unique optical property allows external label-free monitoring of the degradation of the scaffold using NIR, which can penetrate the human skin. When the inverse opal started to degrade, the long-range order began to collapse and the NIR reflection signal became weaker, allowing monitoring of the integrity of the PLA scaffold. However, NIR radiation can only penetrate animal tissues by several millimeters, not enough for most implantation to work. PLA is also commonly co-polymerized with poly(glycolic acid) PGA to form PLGA co-polymer. Interestingly the degradation rate of PLGA can be controlled by varying the PLA:PGA ratio. PLGA inverse opal has also been tested for tissue engineering. In a study on neovascularization, it was found that the characteristic of the vascular networks formed was influenced by the pore size of the inverse opal (Choi et al., 2013). A similar study on bone regeneration was also reported (Choi et al., 2010). Kim et al. demonstrated the preparation of PLGA inverse opal using silica particles as a template (Kim et al., 2018). The inverse opal decolorized in the presence of water and can be used on packaging for food and cosmetics.

In addition to PLA and PLGA, chitosan inverse opal has also been reported. Chitosan is an easy material to manipulate because it can solidify by an increase in the pH to near neutral from an acidic solution. Chen et al. demonstrated the preparation of chitosan inverse opal using silica particles as the template (Chen et al., 2018). The silica template was removed using sodium hydroxide, which does not dissolve chitosan. The materials have been tested for drug delivery while online monitoring for wound healing was also demonstrated. Another commonly used biodegradable polymer for biomedical applications is silk. Kim et al. demonstrated the preparation of a silk inverse opal using PMMA particles as a template, which was removed by solvent solvation using acetone

(Kim et al., 2012). Silk inverse opal has been used for growing stem cells, showing good biocompatibility (Tseng et al., 2017). There are other examples using silk inverse opal materials in biomedical sciences in the literature, including bone regeneration (Sommer et al., 2017) and antibody delivery (Zhang et al., 2020).

The use of biodegradable polymer in inverse opal also has another direction. Some studies use templates that are made from biodegradable polymers. As such, the template can simply be digested under mild conditions (e.g., 37°C in water, with/without the presence of an enzyme) and the inverse opal structure can be easily preserved. For example, biodegradable poly(caprolactone) PCL particles have been used as the template for preparing chitosan inverse opal (Choi et al., 2009). However, in this study, solvent dissolution was performed to remove the PCL template.

4.10 Conclusion

Inverse opal materials are good examples of demonstrating the versatility of nanocasting or nanotemplating technology. These hard templates of polymer particles are much simpler to prepare than those in Chapter 3 and much easier to control than liquid crystal templating as discussed in Chapter 2. In many cases of inverse opal, the design of the materials aimed to increase the surface area and to form a stable material. The enhanced optical property of inverse opals is a welcomed bonus. For many materials, particularly nitrides, the inverse opal pathways may offer a more promising route for increasing the surface area as the collapse of the structure was not as commonly reported. Compared with 2nd or 3rd generation of mesoporous materials, inverse opals are also easier to prepare.

As we also mentioned, the polymer template of inverse opal can be made from biodegradable materials such as poly(lactic acid) and poly(glycolic acid). This is an encouraging development as green syntheses are attractive to many scientific communities. This is a unique feature of inverse opal since calcination or solvent extraction of surfactant templates for mesoporous silica is considered not environmentally friendly. Under the current climate, the possibility of green synthesis would be a distinct advantage.

The superior optical properties also allow them to be potentially used in many sensing applications, notably optical sensing of biomolecules and microbial species. This area is always in high demand as improving public health is on the agenda of many communities worldwide. Sensitive and accurate diagnostic techniques with a fast-responding time are the holy grail for many developments and research in biomedical research. With strong efforts from scientists, inverse opals will have a bright future.

References

Barako, M. T., Sood, A., Zhang, C., Wang, J., Kodama, T., Asheghi, M., Zheng, X., Braun, P. V., Goodson, K. E., (2016). Quasi-ballistic electronic thermal conduction in metal inverse opals, *Nano Lett.*, **16**, pp. 2754–2761.

Blanco, A., Chomski, E., Grabtchak, S., Ibisate, M., John, S., Leonard, S. W., Lopez, C., Meseguer, F., Miguez, H., Mondia, J. P., Ozin, G. A., Toader, O., and van Driel, H. M., (2000). Large-scale synthesis of a silicon photonic crystal with a complete three-dimensional bandgap near 1.5 micrometres, *Nature*, **405**, pp. 437–440.

Chen, J. I. L., Freymann, G., Choi, S. Y., Kitaev, V., and Ozin, G. A. (2006). Amplified photochemistry with slow photons, *Adv. Mater.*, **18**, pp. 1915–1919.

Chen, C., Liu, Y., Wang, H., Chen, G., Wu, X., Ren, J., Zhang, H., and Zhao Y. (2018). Multifunctional chitosan inverse opal particles for wound healing, *ACS Nano*, **12**, pp. 10493–10500.

Chiappini, A., Pasquardini, L., Nodehi, S., Armellini, C., Bazzanella, N., Lunelli, L., Pelli, S., Ferrari, M., and Pietralunga, S. (2018). Fluorescent aptamer immobilization on inverse colloidal crystals, *Sensors*, **18**, p. 4326.

Choi, S. W., Xie, J., and Xia Y. (2009). Chitosan-based inverse opals: Three-dimensional scaffolds with uniform pore structures for cell culture, *Adv Mater.*, **21**, pp. 2997–3001.

Choi, S. W., Zhang, Y., Thomopoulos, S., and Xia, Y. (2010). *In vitro* mineralization by preosteoblasts in poly(DL-lactide-co-glycolide) inverse opal scaffolds reinforced with hydroxyapatite nanoparticles, *Langmuir*, **26**, pp. 12126–12131.

Choi, S. W., Zhang, Y., Macewan, M. R., and Xia, Y. (2013). Neovascularization in biodegradable inverse opal scaffolds with uniform and precisely controlled pore sizes, *Adv. Healthc. Mater.*, **2**, pp. 145–154.

Costacurta, S., Biasetto, L., Pippel, E., Woltersdorf, J., and Colombo, P. (2007). Hierarchical porosity components by infiltration of a ceramic foam, *J. Am. Ceram. Soc.*, **90**, pp. 2172–2177.

Fujishima, M., Sakata, S., Kikoku, M., Ogawa, D., and Uchida, K. (2007). Poly(lactic acid) inverse opal as implantable photonic bandgap materials, *Chem. Lett.*, **36**, pp. 1510–1511.

Gray, B. M., Hassan, S., Hector, A. L., Kalaji, A., and Mazumder, B. (2009). Template infiltration routes to ordered macroporous TiN and SiN_x Films, *Chem. Mater.*, **21**, pp. 4210–4215.

Guo, Q., Zhou, C. Y., Ma, Z. B., and Xueming Yang, X. M. (2019). Fundamentals of TiO_2 photocatalysis: concepts, mechanisms, and challenges, *Adv. Mater.*, **31**, p. 1901997.

Hatton, B., Mishchenko, L., Davis, S., Sandhage, K. H., and Aizenberg, J. (2010). Assembly of large-area, highly ordered, crack-free inverse opal films, *Proc. Natl. Acad. Sci. U. S. A*, **107**, pp. 10354–10359.

Holland, B. T., Blanford, C. F., and Stein, A. (1998). Synthesis of macroporous minerals with highly ordered three-dimensional arrays of spheroidal voids, *Science*, **281**, pp. 538–540.

Jiang, Y., Cui, C., Huang, Y., Zhang, X., and Gao, J. (2014). Enzyme-based inverse opals: a facile and promising platform for fabrication of biocatalysts, *Chem. Commun.*, **50**, pp. 5490–5493.

Jiao, J., Wei, Y., Chi, K., Zhao, Z., Duan, A., Liu, J., and Zheng, P. (2017). Platinum nanoparticles supported on TiO_2 photonic crystals as highly active photocatalyst for the reduction of CO_2 in the presence of water, *Energy Technol.*, **5**, pp. 877–883.

Kim, S., Mitropoulos, A. N., Spitzberg, J. D., Tao, H., Kaplan, D. L., and Omenetto, F. G. (2012). Silk inverse opals, *Nature Photon.*, **6**, pp. 818–823.

Kim, J.-W. Lee, J.-S., and Kim, S.-H. (2018). Biodegradable inverse opals with controlled discoloration, *Adv. Mater. Interfaces*, **5**, p. 1701658.

Lee, W., Kang, T., Kim, S.-H., and Jeong, J. (2018). An antibody-immobilized silica inverse opal nanostructure for label-free optical biosensors, *Sensors*, **18**, p. 307.

Li, J., Zhao, X., Wei, H., Gu, Z.-Z., and Lu, Z. (2008). Macroporous ordered titanium dioxide (TiO_2) inverse opal as a new label-free immunosensor, *Anal. Chim. Acta*, **625**, pp. 63–69.

Liu, J., Liu, G., Li, M., Shen, W., Liu, Z., Wang, J., Zhao, J., Jiang, L., and Song, Y. (2010). Enhancement of photochemical hydrogen evolution over Pt-loaded hierarchical titania photonic crystal, *Energy Environ. Sci.*, **3**, pp. 1503–1506.

Liu, W., Wang, A., Tang, J., Chen, S.-L., Yuan, G., Zhao, K., Li, C., and Liu, X. (2015). Preparation and photocatalytic activity of hierarchically 3D ordered macro/mesoporous titania inverse opal films, *Microporous Mesoporous Mater.*, **204**, pp. 143–148.

Maeng, B., Park Y., and Park, J. (2016). Direct label-free detection of Rotavirus using a hydrogel based nanoporous photonic crystal, *RSC Adv.*, **6**, pp. 7384–7390.

Makuła, P., Pacia, M., Macyk, W. (2018). How to correctly determine the band gap energy of modified semiconductor photocatalysts based on UV-Vis spectra, *J. Phys. Chem. Lett.*, **9**, pp. 6814–6817.

Mandlmeier, B., Szeifert, J. M., Fattakhova-Rohlfing, D., Amenitsch, H., and Bein, T. (2011). Formation of interpenetrating hierarchical titania structures by confined synthesis in inverse opal, *J. Am. Chem. Soc.*, **133**, pp. 17274–17282.

Qi, D., Lu, L., Xi, Z., Wang, L., and Zhang, J. (2014). Enhanced photocatalytic performance of TiO_2 based on synergistic effect of Ti^{3+} self-doping and slow light effect, *Appl. Catal. B: Environ.*, **160**, pp. 621–628.

Seyfoddin, A., Chan, A., Chen, W. T., Rupenthal, I. D., Waterhouse, G. I., and Svirskis, D. (2015). Electro-responsive macroporous polypyrrole scaffolds for triggered dexamethasone delivery, *Eur. J. Pharm. Biopharm.*, **94**, pp. 419–426.

Shen, W., Li, M., Wang, B., Liu, J., Li, Z., Jiang, L., and Song, Y. (2012). Hierarchical optical antenna: gold nanoparticle-modified photonic crystal for highly-sensitive label-free DNA detection, *J. Mater. Chem.*, **22**, pp. 8127–8133.

Sommer, M. R., Vetsch, J. R., Leemann, J., Müller, R., Studart, A. R., and Hofmann, S. (2017). Silk fibroin scaffolds with inverse opal structure for bone tissue engineering, *J. Biomed. Mater. Res. B Appl. Biomater.*, **105**, pp. 2074–2084.

Sordello, F., Duca, C., Maurino, V., and Minero, C. (2011). Photocatalytic metamaterials: TiO_2 inverse opals, *Chem. Commun.*, **47**, pp. 6147–6149.

Subban, C. V., Smith I. C., and DiSalvo, F. J. (2012). Interconversion of inverse opals of electrically conducting doped titanium oxides and nitrides, *Small*, **18**, pp. 2824–2827.

Tetreault, N., Miguez H., and Ozin, G. A. (2004). Silicon inverse opal - A platform for photonic bandgap research, *Adv. Mater.*, **16**, pp. 1471–1476.

Wang, J., Li, Q., Knoll, W., and Jonas, U. (2006). Preparation of multilayered trimodal colloid crystals and binary inverse opals, *J. Am. Chem. Soc.*, **128**, pp. 15606–15607.

Tseng, P., Zhao, S., Golding, A., Applegate, M. B., Mitropoulos, A. N., Kaplan, D. L., and Omenetto, F. G. (2017). Evaluation of silk inverse opals for "smart" tissue culture, *ACS Omega*, **2**, pp. 470–477.

Wang, H., Gu, H., Chen, Z., Shang, L., Zhao, Z., Gu, Z., and Zhao, Y. (2017). Enzymatic inverse opal hydrogel particles for biocatalyst, *ACS Appl. Mater. Interfaces*, **9**, pp. 12914–12918.

You, X., Pikul, J. H., King, W. P., and Pak, J. J. (2013). Zinc oxide inverse opal enzymatic biosensor, *Appl. Phys. Lett.*, **102**, p. 253103.

Zhang, B., Cheng, Y., Wang, H., Ye, B., Shang, L., Zhao, Y., and Gu, Z. (2015). Multifunctional inverse opal particles for drug delivery and monitoring, *Nanoscale*, **7**, pp. 10590–10594.

Zhang, H., Liu, Y., Chen, G., Wang, H., Chen, C., Li, M., Lu, P., Zhao, Y. (2020). Immunotherapeutic silk inverse opal particles for post-surgical tumor treatment, *Sci. Bull.*, **65**, pp. 380–388.

Zhao, Y., Zhao, X., Tang, B., Xu, W., Li, J., Hu, J., and Gu, Z. (2010). Quantum-dot-tagged bioresponsive hydrogel suspension array for multiplex label-free DNA detection, *Adv. Funct. Mater.*, **20**, pp. 976–982.

Zhao, X., Xue, J., Mu, Z., Huang, Y., Lu, M., and Gu, Z. (2015). Gold nanoparticle incorporated inverse opal photonic crystal capillaries for optofluidic surface enhanced Raman spectroscopy, *Biosens. Bioelectron.*, **72**, pp. 268–274.

Zhong, Q., Xie, Z., Ding, H., Zhu, C., Yang Z., and Gu, Z. (2015). Carbon inverse opal rods for nonenzymatic cholesterol detection, *Small*, **11**, pp. 5766–5770.

Chapter 5

Molecular Imprinting: The Key and Lock Match

Humphrey H. P. Yiu, Elizabeth C.H.T. Lau, and Joel Kuhn

School of Engineering and Physical Sciences, Heriot-Watt University, Edinburgh, EH14 4AS, UK

h.h.yiu@hw.ac.uk

5.1 Introduction

In previous chapters, we have already learnt three nanotemplating techniques. They provide important pathways to prepare nanomaterials of many size, shapes and structures. Nonetheless, these are all physical-templating methods, which means that the resultant nanomaterials have no specific chemical properties due to the templates. The chemical properties of these resultant materials are rarely linked to the template once this is removed. If we need to introduce any chemical property, we normally need to introduce specific chemical functional groups after the materials have been made, or simply termed post-synthesis functionalization. *In situ*, or sometimes called one-pot, functionalization is possible but there is no control of the orientation of functional groups. Such specificity is particularly important in biological

Nanotemplating: Engineering Nanocomposite Materials for the Future
Humphrey H. P. Yiu

ISBN 978-981-4316-92-7 (Hardcover), 978-1-003-50999-8 (eBook)
www.jennystanford.com

applications as many biomolecules are defined by both their shape and functional groups.

A technique called molecular imprinting (or MIP) allows us to assemble materials that can have the complimentary chemical functionalities for identifying molecules, not only just according to their size or shape, but to their chemical properties as well. This is another leap from size exclusion we have seen from ordered mesoporous silica and inverse opal, to a more advanced templating on a nanometer scale.

Molecular imprinting refers to the use of a specific molecule as a template to build materials "around" the template molecule, not just by matching the shape but also by matching their chemical functionalities on the surface. Materials of this specificity have already been created by nature. For example, enzymes are known to have a high selectivity towards specific substrates, based on their surface functionalities and shape. Indeed, the specificity of enzymes is so superior that enantiomers (chiral molecules) can be selected and then carry out certain reactions. This property has been exploited in the pharmaceutical industry as chiral molecules are highly valuable in drug synthesis. However, enzymes are still produced and extracted from biological sources rather than from purely synthetic routes. There is a need to explore any possible pathway for synthetic enzymes, or at least materials that can identify specific molecules, but without transformation like enzymes do.

Molecular imprinting was originally aimed to synthesize a material capable of selecting the template molecules from a complex matrix. This idea was first demonstrated as early as the 1930s when Polyakov used silica to make the first synthetic molecularly imprinted material (Polyakov, 1931). In this work, the silica matrices were prepared with additives (or templates) such as benzene, toluene and xylene. After the matrices were dried for 20–30 days, these additives were removed by extensive washing. They were found to have a higher selective adsorption towards the specific additives, i.e., memory for the templates, in case of toluene and benzene. This is highly significant as it is the first breakthrough towards molecular imprinting.

By the 1970s there was another breakthrough when Guenter Wulff and his research team reported their success with preparing a molecularly imprinted material using organic polymer via what is now called the "covalent approach." The polymer prepared could distinguish between the different enantiomers of glyceric acid (Wulff & Sarhan, 1972). Their work expanded well through to the 1980s. Since then, several other breakthroughs have emerged over the following decades in which "non-covalent imprinting" became a widely used technique, which involved the use of other intermolecular forces such as dipole-dipole interactions that arise between molecules and atoms (Sellergren & Allender, 2005).

5.2 General Principles for Molecular Imprinting

In general, molecularly imprinted materials are formed from the assembly of functional monomers around a chosen template molecule followed by polymerization of these monomers. In addition, the formation of these molecularly imprinted polymers, or simply MIPs, may require an optimal amount of cross-linking agent to retain the delicate balance between polymer rigidity and flexibility. After the polymerization reaction, the template with the functional monomers is removed, thus leaving behind the desired mold for recognition. Figure 5.1 presents a graphical illustration on how molecularly imprinted materials are prepared in principle.

MIP recognition works on the basis of intermolecular forces and shape recognition. There are two main approaches to molecular imprinting; the covalent approach, and the non-covalent approach. As mentioned previously, the covalent approach was the first demonstrated by Wulff for molecular imprinting. It involves the formation of covalent bonds between the template molecules and the mold. After the MIP is formed, the template molecules have to be removed from the mold by breaking the covalent bonds. Once the template was removed from the mold, the functionality of the covalent binding sites is left behind and available for binding (or recognizing) target molecules. The main advantage of the covalent approach is the functional groups of the matrix are very selective. However, removal of the template molecule may cause considerable difficulties (needing a powerful reagent to break covalent bonding) as well as the limitations in the types of molecules that could be used as a template, which has to be able to form covalent bonds easily/readily with the functional monomers. Consequently, the non-covalent approach in molecular imprinting has been preferred in many cases.

The non-covalent approach involves the interaction between groups of molecules and monomers, but this does not usually include any covalent bonding; these interactions, termed as template-monomer interactions, are temporary, and reversible. This is particularly important for template molecules in molecular imprinting as the template molecule must be easily removed from the final mold; an advantage over the covalent approach.

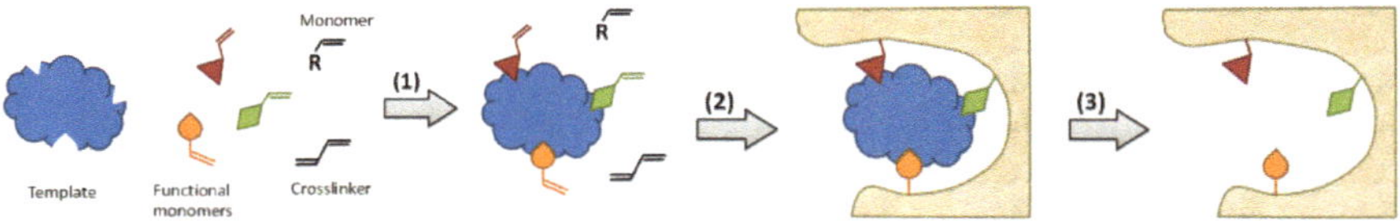

Figure 5.1 The formation mechanism for a molecular imprinting site. Step (1): self-assembly; (2): polymerization; (3) template removal.

5.2.1 Template-Monomer Interactions

Molecular imprinting exploits the various types of interaction between molecules, or more specifically between chemical groups. These interactions can be vaguely categorized into three types; non-covalent interactions, covalent interactions and metal-ligand coordination. (See Fig. 5.2).

5.2.1.1 Non-covalent interactions

Usually, non-covalent interactions refer to the attractive force between groups but no chemical bonding (electron sharing) is formed between these groups. Therefore, these interactions are not permanent in general. Indeed, they are mostly reversible. This can be crucial as the template molecules should be easy to remove after the formation of the imprint. Also, if the imprint is to be used for separations or sensing (see Section 5.4), non-reversible interaction of the imprints and target molecules makes the imprinted materials non-reusable. This is obviously undesirable if the imprinted materials are to be used in separations as the target molecules need to be recovered with simple elution.

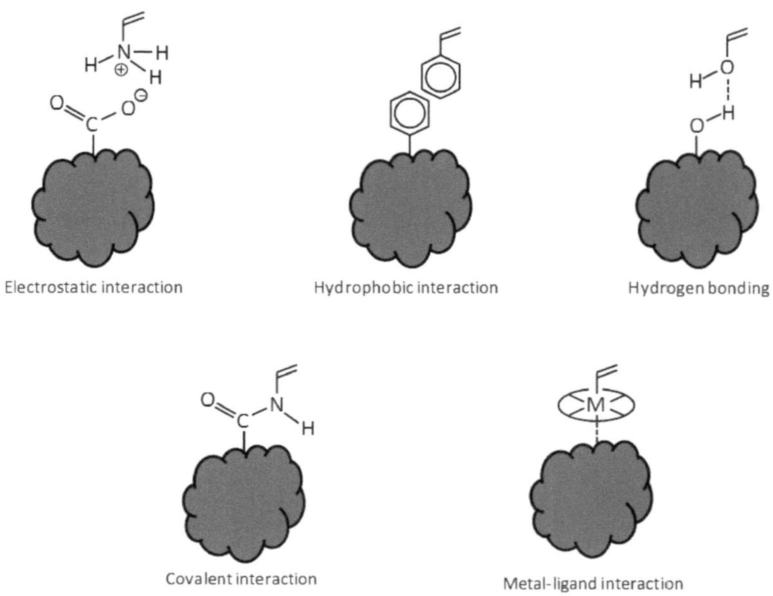

Figure 5.2 The five main types of molecular interactions between template molecules and functional monomers that are commonly used in molecular imprinting technology.

There are three main types of non-covalent interactions are:

(i) Electrostatic interaction

Electrostatic interaction is the attractive force between groups of opposite charge. For example, positively charge groups such as amines ($-NH_3^+$) are attractive to negatively charged groups such as carboxyl ($-COO^-$) groups. This pairing is particularly useful for imprinting biomolecules as many of them have amine groups and/or carboxyl groups. Among amino acids, there are three positively charged amino acids (arginine, histidine and lysine) while two are negatively charged (aspartic acid and glutamic acid).

(ii) Hydrogen bonding

Hydrogen bonding usually refers to the attraction between the partial positive charge of a hydrogen atom on a –OH, –NH or $-NH_2$ group towards the partial negative charge on an "O" or "N" atom. Such partial charge is a result of the significant difference in the electronegativity between neighboring atoms on the hydroxyl and amine groups. Hydrogen bonding is also common in biomolecules, including proteins and nucleic acids.

(iii) Hydrophobic interaction

Hydrophobic interactions are the relationship between two hydrophobic molecules e.g., fats. These interactions arise due to the enthalpy and entropy of the reaction rather than intermolecular forces acting on the hydrophobic fat molecules. For example, fat molecules in water are more likely to agglomerate together than distribute themselves in water. This allows for minimal contact with water and induces strong interactions between hydrophobic molecules.

5.2.1.2 Covalent bonding

As mentioned previously, MIP can be formed by linking the template molecule with a monomer via covalent binding. However, if the covalent method is used, this bonding between the template and the resultant binding cavity must be reversible in order to free up the binding sites. The challenge is that there are not that many reversible covalent bonds that can be used for this purpose. Moreover, the chosen bonding also needs to be compatible with the template molecules. One possible candidate is the disulfide bond (–S-S–), which is reversible upon the availability of redox species (e.g., H_2 and O_2) in the environment. This suggests that both the template molecules and the functional monomer must have a free thiol –SH group, which is not always available. As a result, the covalent method is not commonly used in generating binding sites during MIP synthesis.

5.2.1.3 Metal-ligand interaction

One specific bonding is the metal-ligand interaction. Several organic groups (ligands) such as amines and carboxylic acids can have high affinities towards transition metals by forming coordination bonds via the donation of lone pair electrons to the empty orbitals on the transition metal centers. If the template molecule has an "assessable" transition metal or a transition metal-bound monomer is available, metal-ligand interaction can be used. Upon template removal, an even stronger ligand might be required to disrupt this strong metal-ligand bond. Since the majority of template molecules are transition metal-free and monomers functionalized with transition metals are not widely available, this method is also rarely used.

5.2.1.4 Binding criteria

In order to prepare an imprint able to identify a specific template molecule, a combination of these interactions is required to construct a high degree of specificity. Such a theory has already been demonstrated in nature. For example, the specific binding site of an antibody consists of a sequence of amino acids of various properties. It is generally believed that at least three different types of interactions (e.g., electrostatic, hydrophobic and hydrogen bonding) are required to construct a molecular imprint with a high specificity. It is also more likely that these are chosen from the non-covalent interactions above, due to their wide variety and availability.

5.2.2 Preparation of a Molecularly Imprinted Material

As with most preparations for materials, it should have a target use in mind before designing the materials and their preparation protocol. For molecular imprints, two main considerations have to be taken into account: What is the target molecule(s) and how are the imprints to be used? This decides what the type of MIP material will be, e.g., an organic polymer or a silica-based material. Also, it determines the polymerization procedure to be used.

5.2.2.1 Templates and binding monomers

Of course, the choice of templates is decided by the end use of the imprint. Clearly, if the imprint is aimed to identify and bind to a specific molecule, say myoglobin, the template has to be the target molecule myoglobin. Once that is identified, monomers to be bound onto the template molecule are to be chosen, based on the possible association site with the template. It would be advisable to know the exact chemical structure of the template which can help in choosing the best combination of monomers, with complementary functional groups. In some cases, this can be less straightforward if the exact structure of the template is not known. This is not unusual for some large protein molecules. In such cases, information such as the amino acid sequence could provide some clue on what groups might be on the surface of the molecule for binding.

In addition to the binding groups, the binding monomers also need to have a second part for polymerization, e.g., a vinyl –C=C– group. This has to be compatible with the final base material of the imprint. For example, if the imprint is a vinyl-based polymer, the binding monomer has to have a vinyl group too so that the polymerization will involve these binding monomers. Otherwise, these binding monomers might be washed away with the template molecules during the template removal step.

5.2.2.2 Base imprinting materials and polymerization

Polymers, inorganic or organic, are commonly used as the base material for a molecular imprint. This is because polymers can be formed around the template molecule to provide a certain degree of shape selectivity. Also, the inclusion of binding monomers on the polymer structure will give the ultimate specificity of the imprint. One consideration is the rigidity of the polymer in the medium of use. If the imprint is too rigid, the target molecules can find it difficult to access the binding sites on the imprint. On the other hand, if the imprint is too soft, the shape of the binding site is unlikely to be retained and the shape selectivity is to be lost. Moreover, the position of the binding sites may have too much freedom to move and this can lead to the wrong identification. Since the mechanical property of polymers varies according to the environment (e.g., solvent and temperature), the rigidity of the base material has to be considered in a specific medium. Usually, the flexibility of organic polymers is an advantage over rigid inorganic materials as it allows template molecules to diffuse into and out of the binding sites more easily.

There are a number of forms of MIP polymers for various types of applications. This is to be determined by the nature of the template, the media of use and the interactions chosen. The following are forms of MIP reported in the literature (e.g., Haupt and Mosbach, 1998):

a. polymer beads
b. polymer monoliths (in columns)
c. MIP polymer grafted on silica particles (or other heterogeneous particles)
d. shell-imprinted core-shell particles
e. MIP grafted planar films

Figure 5.3 depicts the designs of these MIP polymers. Because of the vast difference in architecture among these MIP types, the preparation methods for these MIP types are considerably different.

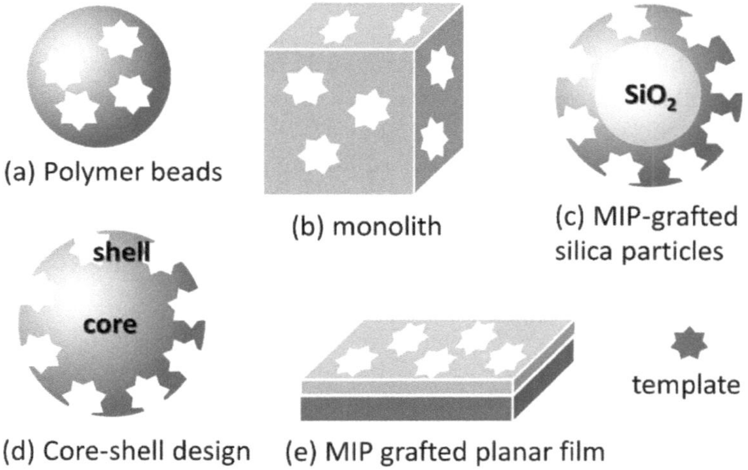

Figure 5.3 Various forms of MIP design.

One of the most common MIPs is in the form of polymer beads, which can be prepared by droplets of polymerization mixture suspended in an aqueous medium, usually with a stabilizer. Micro-emulsion can also be used to prepare MIP polymer beads. An emulsion is to be prepared by suspending an organic polymerization matrix in water with a surfactant and the formation of micelles limits the size of the polymer during growth. This method produces micron-size MIP polymer beads, usually with narrow particle size range. This is also somehow similar to the preparation of inorganic nanoparticles using surfactant templates, as seen in Chapter 1. The size of an MIP polymer can also be controlled by a larger container, such as an HPLC column. This leads to the formation of MIP polymer monoliths in columns if these beads are to be linked chemically.

The other types can be considered as "supported MIP sites on an inactive platform." Choices of these platforms are available, polymer beads, inorganic particles or a planar film. These platform materials can be prepared prior to the MIP formation or even purchased from commercial outlets.

In recent years, technologies have helped to develop MIP with more desirable properties. One notable feature is a higher surface area to volume ratio. An increase in surface area will certainly enhance a higher binding capacity by increasing the number of accessible binding sites per mass or volume of material, leading to a higher efficiency. One strategy is to reduce the particle size of MIP towards nanoparticle synthesis. However, reducing the particle size poses difficulties in downstream separation. To separate MIP nanoparticles, filtration could be difficult with a significant loss on the filters, which could trap small particles effectively. On the other hand, centrifugation is usually laborious and not as effective for smaller particles. Another strategy is to increase the porosity of the MIP polymer, leading to a higher surface area for binding. The pores needed to be large enough for the access/diffusion of target molecules to bind. Porous MIP does increase the binding capacity but, in some cases, it slows down the binding kinetics. This is a compromise that needs to be considered.

5.2.2.3 Template removal

Once the MIP is formed, the next step is to remove the template from the polymers and regenerate the binding sites. An effective removal of the template is very important as this directly affects the binding capacity of the formed MIP. The main principle is to disrupt the interaction between the template molecule and the binding site. Usually, introducing ions, in particular H^+ and OH^- is the most common strategy. For example, strong acid (HCl) and base (NaOH) could be used to remove the template. We also need to ensure that this medium does not destroy the MIP itself. For MAA-based MIP (which is one of the most common type), HCl and NaOH do not cause much damage to the MIP. Salt solution (e.g., NaCl) can also be used when the base polymer is sensitive to pH. However, since NaCl cannot protonate and deprotonate acid and base groups, it is less effective. In some cases, the template may also be destroyed/digested using acid/base/enzymes upon removal. This is only the last resort since some templates can be valuable. Moreover, digested residues may be left in the imprinted sites, reducing the site specificity. Also, MIP could also be used as a pre-concentrating agent. If the template molecules can only be removed by digestion, pre-concentration cannot be carried out. Therefore, the method for template removal must be selected carefully.

5.3 Uses of Molecularly Imprinted Materials

5.3.1 Separation and Chromatography

One of the original objectives of MIP materials is to develop a highly specific method for separating and purifying some structurally complex molecules, e.g., biological enantiomers. Existing methods include some multiple-step separations with each step targeting a specific property of the molecule. This is a time-, labor- and material-consuming process. With the MIP sites in place in the stationary phase, a one-step separation can be developed into an efficient process for the isolation of complex molecules. Generally, the use of MIPs in chromatography can be grouped into several categories as follows:

5.3.1.1 Liquid chromatography

Liquid chromatography is a major branch of separation science and is widely used in the separation of biological samples. Since most biological molecules are thermally sensitive, the use of gas chromatography, which involves the vaporization of analytes, can be completely ruled out. Also, in liquid chromatography, the analytes can be analyzed in their most stable media, such as a buffer with the optimal pH. MIP technology can provide an extra specification to liquid chromatography, in particular shape recognition. Therefore, the use of MIP separation technology coupled with liquid chromatography has a promising future in this area.

There are many successful examples of the separation of biomolecules using a prepared MIP stationary phase. However, the use of MIP in enantiomer recognition is particularly attractive. This is because chiral molecules (enantiomers) are highly valuable in medical/pharmaceutical industries but isolation and purification of chiral molecules can be problematic due to their similarity in chemical properties. For example, the resolution of carbohydrate enantiomers, α- and β-phenyl galactosides, was demonstrated using acetylated β-phenyl galactosides as templates for the preparation of the MIP polymer stationary phase (Nilsson et al., 1995). Another successful example was the resolution of *p*-nitrophenyl α-D-mannopyranoside using a polymer-based MIP stationary phase (Wulff & Vesper, 1978). Amino acids and peptides are another family of chiral biomolecules with plenty of interest. One successful example is the use of phenylalanine (and its derivatives) as a template for the preparation of MIP, using non-covalent imprinting technique. Separation of dipeptides such as Phe-Phe and N-Ac-L-Trp-L-Phe-OMe (with both ends capped) have also been demonstrated previously (Bai et al., 2009). Since the size of these peptide templates increases, the problems associated with the removal of template molecules become more severe. This is due to the decrease in the diffusion rate when the size of a molecule increases. Nonetheless, there are increasingly more successful examples found in the literature, including the separation of ribonuclease A and myoglobin for protein separation using MIP stationary phases (Lin et al., 2006).

Research on using MIP for the analysis of other biological species including antibiotics, anti-cancer agents, anti-inflammatory drugs, central nervous system active agents, and adrenergic stimulants can also be found. Among these biochemicals, steroids are good candidates for MIP preparations because of their rigid structure (multiple cyclic systems) with well-defined, localized functional groups. Examples of using steroids as templates include cortisol, estradiol, and cholesterol (Suda et al., 2017). Steroid-imprinted stationary phase materials can be developed for the use of drug screening. In addition to biological molecules, MIP-based liquid chromatography has also been used for the separation and detection of organic molecules with environmental interest, such as herbicide atrazine and biophenol A.

Despite the vast development of the MIP stationary phase for liquid chromatography found in the literature, a commercialized MIP-based HPLC stationary phase column is yet to be on the market. This is possibly because the MIP technology is a molecule-specific technique, i.e., one MIP stationary phase may only target one or a few molecules. As such large commercial establishments are likely to require a large number of different MIP materials for their operation if they are to exploit the market. That would mean a high production cost but the market may be small. Also, because of their template-orientated preparation, the production cost of MIP materials can be high if the template molecules are expensive and the amount of MIP stationary phase needed could be large. Together with the limited size of the market, commercialization of MIP materials for large-scale separation can be difficult. As a result, in-house prepared MIP materials are used in many MIP-based analyses.

5.3.1.2 Capillary electrochromatography (CEC)

Capillary electrochromatography (CEC) is a combined technique for separating molecules. With the collective advantages of capillary electrophoresis (CE) and liquid chromatography (LC), CEC is a powerful technique for the analysis of charged analytes. The separation of CEC is based on both the retention of a stationary phase (LC) and the electrophoretic mobility due to the applied electric field (CE), so a much higher separation efficiency is observed with CEC compared with LC alone. Also, the associated problem of pressure drop in the HPLC column can be eliminated and the use of a capillary column can be employed to minimize the consumption of stationary phase materials. This is particularly important if the stationary phase materials are expensive to prepare, such as MIP-based materials.

MIP-based CEC has been studied for the resolution of propranolol enantiomers, which can be separated within a 40 s timeframe when a monolithic MIP stationary was used (conventional operation time around 3 h) (Liu et al., 2012). A similar R-propranolol-MIP system can also be used for the resolution of several β-adrenergic antagonists (Nilsson et al., 1997). Other examples include using an S-ropivacaine-MIP for the separation of local anesthetics (Karlsson et al., 2001), L-phenylalanine anilide-

MIP for the resolution of phenylalanine (Halhalli et al., 2015), and (+)-ephedrine-MIP for the separation of ephedrine and salbutamol enantiomers (Tian et al., 2013).

5.3.1.3 Extraction

One of the major areas of research on MIP materials is their use for solid-phase extraction. Although MIP has been used in liquid chromatography (LC and CEC), it can still be difficult to detect a specific molecule in an analyte with a rather complex matrix, such as a sample from an animal source, due to interference between molecular species. This can made even more challenging if the target molecule is of a very low concentration, depending on the detection limit of the detector of the system. As a result, pre-concentration of analytes may be very useful not only to remove the interfering species but also to increase the concentration of the target molecule to a detectable level. The bound molecules were then eluted from the MIP for quantification. In this way, the concentration of analytes can be increased 100-fold. Also, this can be used as a purification process for chemicals or biochemicals.

Before extraction with MIP, usually a pre-extraction is carried out on the samples by using standard solid-liquid or liquid-liquid extraction. Then the MIP material is added to selectively bind the target molecules. These MIP binding materials can be of particle, film, or monolithic form. Such an extraction technique has been used to analyze pollutants (including chlorotriazine pesticides and phenylurea herbicides) and biological species (tamoxifen, phenytoin, quercetin, and 7-hydroxycoumarin) from human samples (blood plasma or urine). These samples usually have a very complex composition and the amount of the target molecules is often low for standard liquid chromatography, even with the aid of MIP technology.

5.3.2 Sensors

Incorporation of molecule-targeted MIP materials into an electro-sensing device will significantly enhance the sensitivity and specificity of sensors. This concept has been explored and many reports on MIP-based sensors can be found in the literature (Guo et al., 2020). There are several types of MIP-based sensors and we discuss a number of selected techniques based on their sensing principles.

5.3.2.1 Fluorescence and chemiluminescence sensing

Fluorescence spectroscopy has a known sensitivity down to a sub-μM level and therefore it is useful for the detection of species at a low concentration in a matrix. Two main types of MIP-based fluorescence sensing can be found in the literature: fluorescent MIP materials and fluorescent target species (Wan et al., 2016). The former technique is based on measuring the change in fluorescence signal of an MIP material before and after the binding of the target molecule. On the other hand, the latter one relies on the fluorescence signal from the target species binding onto a non-fluorescent MIP material, which is non-fluorescent. The principle of these two approaches is shown in Fig. 5.4.

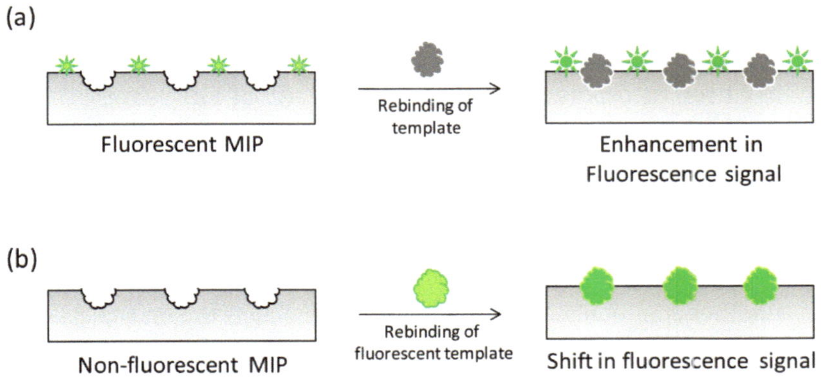

Figure 5.4 Two strategies for fluorescence sensing using MIP, (a) using a fluorescent MIP and (b) sensing a fluorescently labeled substrate.

Fluorescent MIP materials can be prepared by post-synthesis grafting and *in situ* grafting. For post-synthesis grafting, the MIP material is prepared with a chosen method (as described in Section 5.2) and followed by the grafting of fluorescent species onto the MIP material. For example, a sialic acid-imprinted polymer was treated with ophthalaldehyde, a fluorescent agent and fluorescence signal of this MIP polymer was found to be enhanced upon rebinding of sialic acid (the template) (Piletsky, et al., 1996). Such a technique has also been used in preparing MIP materials for detecting triazine herbicides and the antibiotic chloramphenicol. Fluorescent reagents can also be introduced during the preparation of MIP polymer, with a designed fluorescent monomer to be copolymerized with the base monomers. This approach has been applied to the detection of specific carbohydrates (cyclic guanosine monophosphate) and amino acids (L-tryptophan) (Liao et al., 1999).

Detections of fluorescent species on an MIP polymer have also been reported. Matsui et al. studied the binding of alkaloids (cinchonidine and cinchonine) onto a trifluoromethacrylic acid-functionalized polymer and a shift in the fluorescence signal was observed upon binding of the template molecules. Detection of other fluorescent targets studied using MIP technology includes flavonol and propranolol (Matsui et al., 2000).

Chemiluminescence-based MIP materials for sensing have also been studied, but not as common as fluorescence-based studies. For example, europium, a commonly used luminescent element, has been incorporated into the binding site of MIP materials for the detection of organophosphate insecticides and herbicides. The detection limits for these pollutants were down to an impressive parts-per-trillion (ppt) range (Southard et al., 2007).

5.3.2.2 Sensing for organics in small quantities

Another commonly used technique for sensing is the quartz crystal microbalance (QCM) device. By incorporating a thin layer of MIP polymer on the quartz crystal plate, the rebinding of substrate molecules will give a signal through mass-sensitive transduction. There are many different designs of such a QCM-MIP sensing device and Fig. 5.5 illustrates a generic design with several common features. QCM-MIP sensing systems have been used to detect cyclodextrins, calixarenes and paracyclophanes. One area of specific interest is the detection of glucose using QCM-MIP sensors. A high demand for precise but simple glucose-sensing devices was driven by the increasing cases of diabetes globally. This was successfully demonstrated using an MIP poly(o-phenylenediamine) film. (Ersöz et al., 2005).

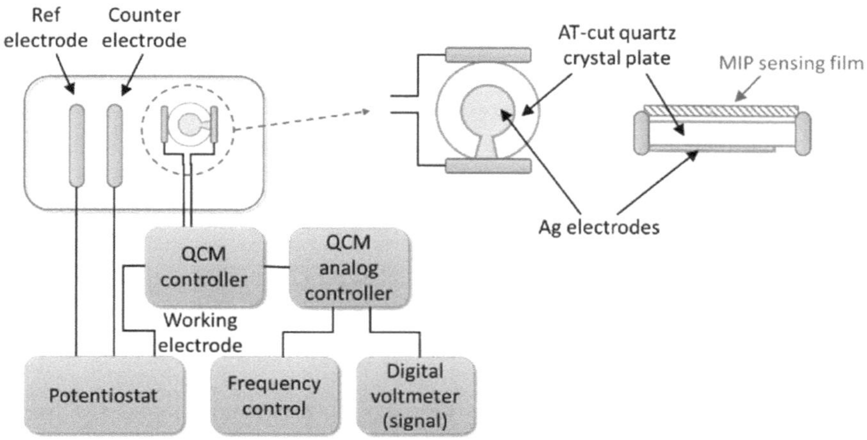

Figure 5.5 The basic construction of a sensing device using combined quartz crystal microbalance (QCM) with MIP (adapted from Ersöz et al., 2005).

Instead of mass-sensing, the rebinding of substrates onto the MIP polymer can also be monitored by changes in electro-conductivity of the MIP layer. During binding, there are changes in the permeability of the MIP membrane and such a phenomenon has been used in the detection of theophylline. A similar system was developed for sensing morphine. In general, an MIP sensing electrode can be prepared by spin coating the MIP polymer onto an electrode. By mixing the MIP polymer with carbon paste, an electrode can also be constructed with MIP sensing capability. The detection limit is usually fairly good; it can be down to a sub-ppb level (Hattori et al., 2004). This concept can be used to develop sensing devices for narcotics.

Surface Plasmon resonance spectroscopy (SPR) has also been used with MIP technology for sensing. SPR measures the shifts in the resonant angle of incidence, due to the changes in the dielectric properties of a film, which can be an MIP polymer film. Such changes in dielectric properties can be triggered by the rebinding of molecules onto the MIP sites on the film. An SPR-MIP system has been developed for the detection of theophylline, zanthine and caffeine. Detection of other biomolecules including adenosine has also been demonstrated (Pernites et al., 2010).

5.3.3 Catalysis and Synthesis

Enzymes can be considered as one of the most fascinating molecular engineering systems in nature as these biocatalysts can select certain molecules to carry out some specific reactions and give specific products. Such a high selectivity is even capable of "choosing" chiral species, which are highly important in biology. This high selectivity in shape is created by the unique sequence of amino acids, combination of surface chemistry and foldings in structure. On the other hand, synthetic catalysts are still somehow short of shape selectivity, particularly on chiral species. Enzyme mimics with high enantioselectivity is the ultimate goal of many research works in this area. Moreover, thanks to the specific engineered site to stabilize the intermediate, the activation energy (E_a) of the reaction is usually much lower than the non-catalytic transformations. Hence, enzymatic catalysis can be carried out at low temperatures (<40°C) and pressure (1 atm), while other systems cannot always achieve this. Lower reaction temperature and pressure can translated into lower costs of manufacturing due to savings from energy consumption. Again, this gives another push for the need for enzyme-mimicking systems.

The principle of molecular imprinting technology allows us to create similar systems to enzymes, with high selectivity of reactants according to their shape and surface chemistry. Such a concept gives scientists some possibilities to engineer the ultimate enzyme mimics, the desired catalysts for the manufacture of fine chemicals. However, the approach to designing a catalyst using MIP technology is rather different from that for the separation and detection of substrates. This is because many reactions involve more than one reactant or product. Therefore, MIP catalysts are designed to stabilize the reaction intermediates, or *transition state analogue* (TSA), as shown in Fig. 5.6. By doing so, the E_a can be lowered and therefore the rate of reaction can be increased. We need to remember that TSAs do not always represent the intermediates of lowest E_a. Nonetheless, the goal here is to create a synthetic pathway of lower overall E_a, and enhanced reaction rate, ideally with a higher selectivity.

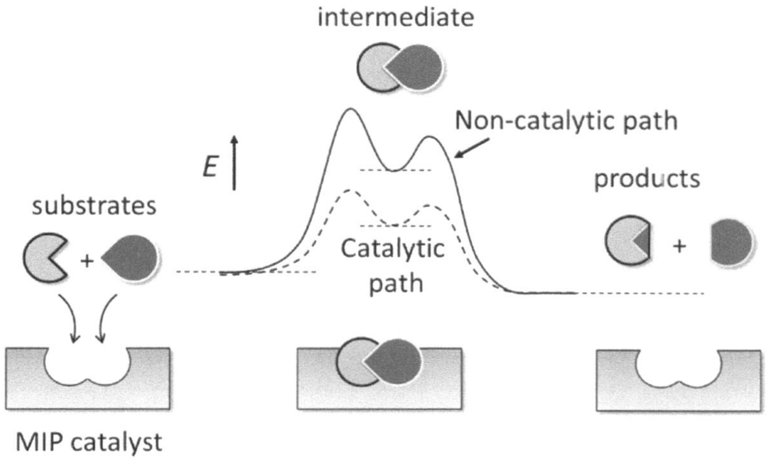

Figure 5.6 Illustration showing the energy profile of a reaction via an MIP catalytic pathway compared to a non-catalytic pathway.

5.3.3.1 Hydrolase-type catalysts

There are many categories of enzymes/enzymatic reactions that are useful in chemical industries. Hydrolysis plays an important role in biological systems; digestion involves the hydrolysis of carbohydrates, proteins and fats. Also, the reverse reaction, water elimination, is a key building mechanism for proteins, esters (e.g., fats), carbohydrates and many other biomolecules. Hydrolases, responsible for hydrolysis, are ubiquitous in our body, namely amylase from saliva, lipase from pancreatic juice and trypsin. Enantioselective hydrolysis (or esterification, the reverse reaction) has been used for the dynamic kinetic resolution, DKR, synthesis of chiral alcohols. DKR is an important reaction for the production of chiral ingredients, which are highly sought after for the synthesis of pharmaceutical products. Such reactions have been dominated by the use of lipases, which catalyze both esterification and hydrolysis of esters. Hence, hydrolase mimics can be of high value to industries.

For example, MIP polymers prepared using phosphonate ester transition state analogue (TSA) were introduced in a review by Wulff (Wulff, 2002). These polymers were capable of catalyzing the hydrolysis of di-aryl esters, and surprisingly, carbonates and carbamates at higher rates. The rate of hydrolysis of carbonates and carbamates using these phosphonate ester TSA-imprinted polymers has been enhanced by up to 3000-fold when compared to non-catalyzed reactions. Another MIP polymer catalyst was reported using amidine and imidazole monomers, both acting as arginine and histidine mimics (Chen et al., 2002). This work moves MIP catalysts one step further towards artificial enzymes.

5.3.3.2 Redox catalysts

Besides hydrolases, redox enzymes are also common in biological systems. These catalysts allow oxidation and reduction reactions to transform organic molecules. This is one mechanism to protect our body from toxins by transforming these toxic molecules into some less toxic forms. Redox enzymes (oxidoreductases) usually possess catalytic sites with a transition metal center, which is mainly responsible for the redox reaction. For example, cytochrome c, a known redox catalyst found in the heart of animals, has an iron complex in its active site. The peptide structures around the transition metal center are to offer the selectivity on shape and surface chemistry of the substrates, boosting the rate of reaction and lowering the activation energy E_a.

Although transition metal complexes are known to be capable of redox catalysis, unfortunately, they tend to have low shape selectivity. Logically, if these metal complexes are to be incorporated into MIP sites on a polymer support, a higher shape selectivity, particularly chiral selectivity, towards substrates may be achieved. MIP catalysts with a Pd, Pt or Rh as the transition metal center have been reported. For instance, an MIP polymer with a chiral polymerizable Mn complex has been reported for the epoxidation of alkenes (Jacobsen-type catalysts) (Disalvo et al., 2002). This MIP catalyst can catalyze the epoxidation of styrene using m-chloroperbenzoic acid. Chiral oxidation of the styrene analogue can give benzylic alcohols, which are precursors for many pharmaceutical products. Another example using Rh-MIP and Cu-MIP for the hydrogenation of enamides showed that the nature of the metal center influenced the chirality of the product, with Rh-MIP producing predominately L-enantiomers while Cu-MIP giving the opposite (Lee et al., 2009).

5.3.4 Pseudo-Immunoassays

Immunoassays have been widely used in biotechnology for the detection of specific biomolecules. However, there is a need for replacing the use of antibodies for such detections due to the disadvantages associated with antibodies. MIP materials do have some structural and functional similarities to antibodies; both have a specific target molecule based on their binding preferences. The origin of these two types of species, however, cannot be more different. Table 5.1 summarizes the advantages and disadvantages of using antibodies and MIP materials for the detection of biomolecules.

As for pseudo-immunoassays, MIP has been used for the detection of theophylline and diazepam using ^{3}H-labelled molecules (Bedwell & Whitcombe, 2016). Detection of fluorescein-labeled trizaines using triazine and atrazine-MIP was also demonstrated.

Table 5.1 Advantages and disadvantages of using antibodies and MIP for the detection of biomolecules

	Advantages	Disadvantages
Antibodies	• A large selection of known antibodies	• Unsuitable to be used in organic solvents
	• Well-established techniques	• Use of animals for production
	• High specificity	• High associated cost
		• Non-recyclable
MIP materials	• Can be used in organic solvents or aqueous media	• Specific MIP to be made with specific template for each assay
	• Higher stability	• Techniques still to be refined
	• Synthetic, no animals required	• A distribution of affinity for the target compounds
	• Recyclable	

5.3.5 Drug Binding and Delivery

Although the development of MIPs is still mainly directed towards a variety of analytical applications, such as sensors and separation technologies, one application under the spotlight in the scientific community is their uses for therapeutic purposes as advanced drug delivery systems. Developing more efficient techniques for delivering drugs has been sought after for many years due to the inefficiencies experienced in current administration methods with the lack of targeting and controlled delivery modes. MIPs have been considered as a solution to this problem. The MIPs prepared from the non-covalent approach is implemented for this type of research, as they are less densely crosslinked. Many of the monomers used to prepare these materials are water soluble and can undergo swelling when influenced by an external stimulus (e.g., thermal changes). This mechanism enables the release of drugs from the MIP-drug complexes, but some obstacles have to be tackled first before this can be achieved.

Six main criteria have to be considered before MIPs can be used in drug delivery systems:

1. Size of the overall MIP particles (this includes both the template and the matrix)
2. Selectivity of MIPs
3. Ensuring the drug will get to the chosen destination (i.e., targeting)
4. Release mechanism to deliver the desired drug
5. Toxicity of the polymer matrix
6. Removal of the polymer from the body (if not exuded naturally)

Table 5.2 summarizes examples of MIPs from the literature that have the potential to be used in drug delivery systems. Entries 1–10 are examples of using “model” drugs, which can be delivered by traditional methods, e.g., orally. Therefore, these examples are only for proof-of-concept. Another reason for the choice of “model” drugs is due to their low cost. On the other hand, entries 11–14 on the table are the highly sought-after drugs for cancer or antibacterial treatments and have been used in research on drug delivery. Both cancer treatment and antibacterial areas are significant issues in society today.

Table 5.2 Examples of MIPs from literature that have the potential to be used in drug delivery systems

	Template molecule	Use of drugs	Functional monomer	Cross linker	Ref.
1	Propranolol	Beta Blocker for high blood pressure, heart dysrhythmias, angina, or previous heart attacks	• MAA, MMA, NAA • MAA, 4-VPA, • MAA • MAA, HMOA	EGDMA TRIM DVB	(a)
2	Sulfasalazine	Arthritis, Irritable Bowel Syndrome, Crohn's Disease	• MAA	EGDMA	(b)
3	Timolol	Beta Blocker for hypertension, migraines and glaucoma	• MAA, MMA, HEMA • MAA, DEAA • MAA, ITA, DEAA, HEMA, SiMA, DMAA • MAA, MMA	EGDMA	(a, c)
4	Theophylline	Chronic obstructive pulmonary disease	• MAA	EGDMA	(a)
5	Ephedrine	Asthma and Hay fever relief	• TFMAA	DVB, EGDMA	(d)
6	Citalopram	Antidepressant	• MAA	TRIM	(e)
7	H1-antihistamines	Negating the Histamine effects	• HEMA	EGDMA	(f)
8	Ibuprofen	Non-steroidal anti -inflammatory drug	• MAA	EGDMA	(a)
9	Morphine	Narcotic Analgesic drug	• HEMA	EGDMA	(g)
10	Paracetamol	Analgesic drug	• MAA, 4-VPY	EDGMA	(a)

(*Continued*)

Table 5.2 *(Continued)*

	Template molecule	Use of drugs	Functional monomer	Cross linker	Ref.
11	5-Fluorouracil	Anti-Cancer Drug	• MAA	EDGMA	(b)
12	Norfloxacin	Broad Spectrum Antibiotic	• AA, 4-VPY	EDGMA	(h)
13	Levofloxacin	Broad Spectrum Antibiotic	• PY	EDGMA	(i)
14	Paclitaxel	Anti-Cancer drug	• PEG-FA	EDGMA	(j)

Abbreviations: MAA, Methacrylic acid; MMA, Methyl methacrylate; NAA, N-naphthylacrylamide; 4-VPA, Vinyl phosphonic acid; HMOA, 2-Hydroxypropyl methacrylamide; EGDMA, Ethylene glycol dimethacrylate; TRIM, trimethylolpropane trimethacrylate; DVB, Divinylbenzene HEMA, 2-hydroxyethyl methacrylate; DEAA, N,N-Diethyl acrylamide; SiMA, 1-(tristrimethyl-siloxysilylpropyl)-methacrylate; TFMAA, 2-(Trifluoromethyl)acrylic acid; 4-VPY, 4-Vinylpyridine; AA, Acrylic Acid; PY, Pyrrole; PEG-FA, PEG-folic acid.

References: (a) Hilt & Byrne, 2004; (b) Puoci et al., 2004; (c) Nostrum, 2005; (d) Piletsky et al., 2002; (e) Abdouss et al., 2011; (f) Venkatesh et al., 2008; (g) Anderson et al., 1995; (h) Alvarez-Lorenzo et al., 2006; (i) Mazzotta et al., 2012; (j) Esfandyari-Manesh, et al., 2016.

Regarding cancer treatments, most anti-cancer drugs are administrated through the body; they have no "direct" target and cause damage to the healthy cells/tissues that come into contact with them. We must remember that most anti-cancer drugs are invasive and carry some certain degree of cytotoxicity in order to kill cancer cells. Moreover, some cancers can be located in deep tissues in the body and thus cannot be reached easily (e.g., bone, brain cancers or lung-related problems). These often require surgery, which always comes with increased risks. Antibiotics are another problem in society today, as more bacteria are becoming resistant to antibiotics, as described as "antimicrobial resistance" or AMR. This due to poor management and use of antibiotics, including misuses in agriculture and by patients who do not take the correct dosages, as well as antibiotic pollutions in the aquatic environment. Both these are perfect examples of how MIPs may be able to be used to overcome the problems that have arisen with these drugs. Figure 5.7 illustrates the basic principle of drug purification and delivery using MIPs.

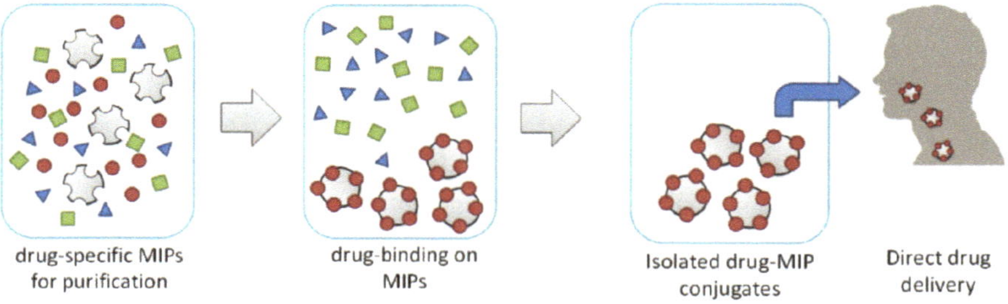

Figure 5.7 Illustration showing the principle of combined drug purification and delivery using MIP.

5.3.5.1 Drug targeting

To improve drug delivery efficacy, scientists have developed targeted delivery, so as not to waste valuable drugs on healthy tissues/organs, where the drug may indeed be harmful. However, this is not straightforward to achieve in an *in vivo* environment. Ensuring that the MIP-drug complex could localize to its target destination (infected tissues or organs) is also one challenge for the development of MIPs in drug delivery in recent years. It has been extrapolated how MIPs can mimic enzymes but it has also been claimed that they can mimic antibodies (Nostrum, 2005). Such speculation has already been discussed in pseudo-immunoassays previously. This property would be very useful in targeted delivery by targeting specific cells/cell types in the body, using the characteristic receptors on cells and tissues. Liposomes and antibodies have already been reported as drug carriers and targeting devices, so a similar design for MIPs should be possible to carry targeted delivery properties (Nostrum, 2005). The fact that most tissue cell types in the body have specific markers or receptors on the surface allows differentiation between cell types. Depending on the nature and chemistry of these markers, MIPs can be designed to have a geometry, as well as complimentary chemistry, that will fit the markers or receptors. Such property can mimic a targeting vector (e.g., antibodies) and allow the MIP to specifically target a certain cell type (Sellergren & Allender, 2005). Another key criterion for a successful delivery system would be controlled drug release or dosage mechanism using MIPs; this will be discussed in the next subsection. Ultimately, the application is steered towards drug delivery using MIPs *in vivo*, the MIPs would be better to be prepared in an aqueous environment for better biocompatibility by minimizing the use of organic solvent. (Puoci et al., 2007) One difficulty is that the surface of the MIP can adsorb unwanted biomolecules such as peptides and lipids, leading to inefficiencies in targeting, by masking the targeting sites. This is a natural defense mechanism of our human body to defend foreign "intruders." Currently, these targeting applications of MIP are still needing improvement for *in vivo* applications, such as enhancing the safety for human consumption, but certainly a research direction in the future.

5.3.5.2 Controlled drug release

Another aspect of a successful drug delivery system is to have precise control over the rate of release of drugs from the delivery agent, no matter whether this is an MIP or other solid-state platform. There are several mechanisms for controlled release being proposed by scientists but their practicality can be sometimes questionable. "Intelligent" release mechanisms are highly sought after in the field of using MIPs for drug delivery. These mechanisms for drug release are based on the changes in environmental parameters such as pH, temperature, ion concentrations or even the availability of an enzyme. However, the main objective is to develop a mechanism that can release the drug from the MIPs in a dosage form, instead of a patient taking doses themselves. Using an MIP to control the dosages could lead to many potential benefits for patients who are undergoing intensive treatments, including maintaining drug levels for treatment, reduction in side effects to the patient and reduction in treatment visits, which posts economic benefits to healthcare organizations.

In this area of MIP development, a non-covalent approach has been the preferred method for producing MIPs for use in controlled drug release due to MIP's flexible structures. Non-covalent approaches also tend to be more favorable in terms of binding and release kinetics (Cunliffe et al., 2005). It is also worth mentioning that to have better control of drug delivery, a uniform spherical shape for the MIPs is desirable. If the MIP were to be non-uniform, irregular swelling of the MIP will occur due to their geometry being significantly deviated from a spherical shape (Puoci, et al., 2010). Currently, in most studies, a stimulus is required to release the drug from the MIP-drug complex. These stimuli can be changes in pH, changes in temperature, ionic strength, electric fields and irradiation by UV or visible light (Cunliffe et al., 2005). Unfortunately, not all these release stimuli can be applicable in an *in vivo* environment. Table 5.3 summarizes the use of these stimuli for drug delivery *in vitro* and *in vivo*.

Table 5.3 Comparisons of various stimuli modes for drug release from MIP

Stimuli	Working principles	*In vitro* application	*In vivo* application
pH	Protonation or deprotonation of polymer functional groups can cause swelling or collapse in polymer structure. The swelling action in MIP triggers release of drug MIP.	Changes to pH *in vitro* (e.g., cell culture) are easily achieved.	Cancer cells are usually slightly more acidic (pH 5–6) than the surrounding cells.
Temperature	Thermosensitive polymers exhibit a critical solution temperature (LCST), where the polymers become water-soluble, hence triggering drug release from MIP polymer when temperature < LCST.	Wide temperature range can be tested *in vitro*.	The body's optimum temperature range is 36.1–37.2°C, changes beyond this range can cause harm to humans.
Ionic strengths	Changes in ionic strength affects the structure of the MIP polymer, causing swelling to trigger release of drug.	Ionic strengths can be manipulated *in vitro* by altering the composition of culture medium.	High concentrations of ions in blood can be used to release drug from the MIP. However, this might not be always practical due to ion regulation by kidneys (Blantz et al., 2007).

(*Continued*)

Table 5.3 (*Continued*)

Stimuli	Working principles	*In vitro* application	*In vivo* application
Electric fields	Changes in charge distributions can cause swelling of some polymers, hence inducing drug release.	Electric fields can easily be implemented *in vitro* by inserting electrodes in cell culture.	Unlikely to generate an electric field inside the body. Generating electric fields can interfere with heart functions.
UV light	Some polymer networks can absorb UV radiations and swell to release drug, due to increase in energy in polymer.	Easy to simulate for drug release using external UV sources.	UV radiation cannot penetrate far enough through the skin for stimulation. UV also causes abnormal cell growth, resulting in tumors. Not suitable for *in vivo* applications.
Visible light	Some polymer networks can absorb radiation from visible light. Increase in energy allows the polymer to swell and release drug.	Easy to simulate for drug release using external light sources.	Visible lights are unlikely to penetrate through the dermal layer.

5.4 Magnetic MIP (MMIP)

Molecularly imprinted polymers can also be incorporated with magnetic properties for enhanced separation/recovery using an external magnetic field. These "magnetic MIPs" or simply MMIPs can be prepared in two main pathways; coating MIPs on a bulk magnetic material or polymerizing the MIP with functionalized magnetic nanoparticles. Figure 5.8 shows the basic designs of MMIPs. The most common one is a core-shell particle design, in which the magnetic components are coated with the polymer matrix. In some cases, a porous polymer network embedded with magnetic nanoparticles can also be formed. In some special cases, when the MMIP was required to be conductive, magnetic carbon nanotubes (multi-walled or single walled) or magnetic graphene oxides are used instead. This type is mainly used in electrochemical detection of organics using electrodes.

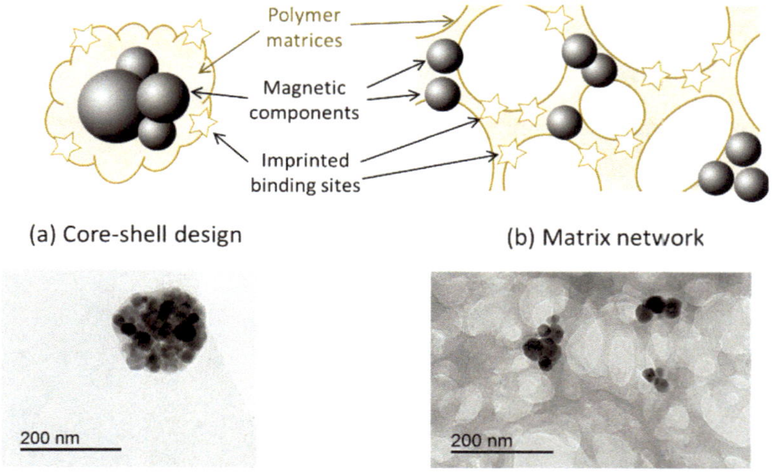

Figure 5.8 Illustration (upper panel) and TEM images (low panel) of (a) a core-shell MMIP particle and (b) a MMIP network.

MMIPs can enhance the recovery of imprinted materials after binding the target molecules from a sample. Although non-magnetic MIP materials can be separated from the liquid samples using filtration or centrifugation, these two methods would not work well if the sample itself contains solid debris or solid particulates. This can be very common in samples from the environment such as sewage or water samples from rivers and sea. As such these unwanted solid materials will also be filtered out with the precious MIPs, and the mixture will now be difficult to separate. If the MIPs carry a good magnetic property, they can be recovered by a magnetic field without filtration or centrifugation, providing the other solid particulates are not magnetic. Figure 5.9 illustrated how this can work with MMIPs.

There are many examples of MMIPs (Aylaz et al., 2021) and they have shown that they are capable of selective binding in complex matrices. Applications of MMIPs have been seen in environmental, food, and medical applications. For example, Kuhn et al. demonstrated the synthesis of MMIP using two antibiotics (erythromycin and ciprofloxacin) as templates (Kuhn et al, 2020).

A vinyl-functionalized magnetic nanoparticle was used as the magnetic component while the vinyl functionality on the surface can be crosslinked with the monomers. It was found that these two MMIPs can selectively bind the target antibiotic molecules from both methanol-water mixture and milk matrices. Such examples highlight the use of MMIPs for selective binding and extraction in complex samples.

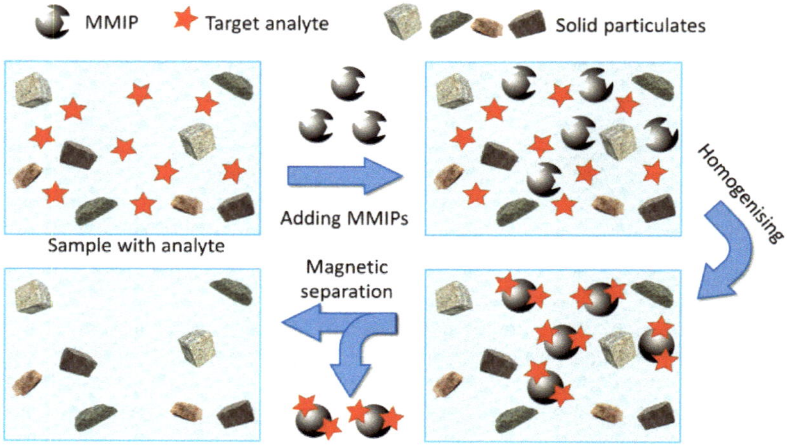

Figure 5.9 Illustration showing how MMIPs work in a complex sample matrix with solid particulates.

5.5 Molecularly Imprinted Materials: The Future Outlook

Similar to other areas of materials chemistry, as the recognition is becoming more specific, MIP materials are getting more complex. Although the preparation methods of MIPs have been diversified and many methods were developed to tailor certain applications, the principle of MIP binding is largely the same. Despite the high selectivity towards particular substrates, MIPs always suffer from their own merit as being "too specific" for a small number of targets. As a result, the use of MIP materials on a commercial scale is still in initial development, although it has started to grow in recent years.

On the other hand, the development of MIPs has already been moving from imprinting some testing templates as proofs of concept into MIPs for specific molecules, such as drugs, proteins, and steroids. Their ability to identify the substrate is improving through using different types of polymers, providing various physical properties to serve for the identification. One property of MIPs that has to be improved is their reusability. Since they are rather costly to prepare, it is only economical if the materials can be reused without losing their specificity, or if the application is costly, such as medical applications.

Another development that can be seen is that MIPs are to be incorporated into designed devices, as a small but essential part for identification. For example, MIP can be used for identifying certain antibodies in a blood testing device. This area is also growing due to the expanding demand in both the medical sector and the society.

References

Abdouss, M., Asadi, E., Azodi-Deilami, S., Beik-Mohammadi, N., and Aslanzadeh, S. (2011). Development and characterisation of molecularly imprinted polymers for controlled release of citalopram, *J. Mater. Sci.: Mater. Med.*, **22**, pp. 2273–2281.

Alvarez-Lorenzo, C., Yanez, F., Barreiro-Iglesias, R., and Concheiro, A. (2006). Imprinted soft contact lenses as norfloxacin delivery systems. *J. Control. Release*, **113**, pp. 236–244.

Anderson, L. I., Muller, R., Vlatakis, G., and Mosbach, K. (1995). Mimics of the binding sites of opioid receptors obtained by molecular imprinting of enkephalin and morphine, *Proc. Natl. Acad. Sci. U. S. A.*, **92**, pp. 4788–4792.

Aylaz, G., Kuhn, J., Lau, E. C. H. T., Yeung, C.-C., Roy, V. A. L., Duman, M., and Yiu, H. H. P., (2021). Recent developments on magnetic molecular imprinted polymers (MMIPs) for sensing, capturing, and monitoring pharmaceutical and agricultural pollutants, *J. Chem. Technol. Biotechnol.*, **96**, pp. pp. 1151–1160.

Bai, L., Sheeley, S., and Sweedler, J. V. (2009). Analysis of endogenous D-amino acid-containing peptides in metazoa, *Bioanal Rev.*, **1**, pp. 7–24.

Bedwell, T. S., and Whitcombe, M. J. (2016). Analytical applications of MIPs in diagnostic assays: future perspectives, *Anal. Bioanal. Chem.*, **408**, pp. 1735–1751.

Blantz, R., Deng, A., Miracle, C., and Thomson, S. (2007). Regulation of kidney function and metabolism: a question of supply and demand, *Trans. Am. Clin. Climatol. Assoc.*, **118**, pp. 23–43.

Chen, W., Han, D.-K., Ahn, K.-D., and Kim J.-M. (2002). Molecularly imprinted polymers having amidine and imidazole functional groups as an enzyme-mimetic catalyst for ester hydrolysis, *Macromol. Res.*, **10**, pp. 122–126.

Cunliffe, D., Kirby, A., and Alexander, C. (2005). Molecularly imprinted drug delivery systems, *Adv. Drug Deliv. Rev.*, **57**, pp. 1836–1853.

Disalvo, D., Dellinger, D. B., and Gohdes, J. W. (2002). Catalytic epoxidations of styrene using a manganese functionalized polymer, *React. Funct. Polym.*, **53**, pp. 103–112.

Ersöz, A., Denizli, A., Özcan, A., and Say, R. (2005). Molecularly imprinted ligand-exchange recognition assay of glucose by quartz crystal microbalance, *Biosens. Bioelectron.*, **20**, pp. 2197–2202.

Esfandyari-Manesh, M., Darvishi, B., Ishkuh, F. A., Shahmoradi, E., Mohammadi, A., Javanbakht, M., Dinarvand, R., and Atyabi, F. (2016). Paclitaxel molecularly imprinted polymer-PEG-folate nanoparticles for targeting anticancer delivery: Characterization and cellular cytotoxicity, *Mater. Sci. Eng. C Mater. Biol. Appl.*, **62**, pp. 626–633.

Guo, X., Li, J., Arabi, M., Wang, X., Wang, Y., and Chen, L. (2020). Molecular-imprinting-based surface-enhanced Raman scattering sensors, *ACS Sensors*, **5**, pp. 601–619.

Halhalli, M. R., and Sellergren, B. (2015). Insights into the formation, structural properties and performance of RAFT polymerized L-phenylalanine anilide molecularly imprinted polymers, *Polym. Chem.*, **6**, pp. 7320–7332.

Hattori, K., Hiwatari, M., Iiyama, C., Yoshimi, Y., Kohori, F., Sakai, K., and Piletsky, S. A. (2004). Gate effect of theophylline-imprinted polymers grafted to the cellulose by living radical polymerization, *J. Membr. Sci.*, **233**, pp. 169–173.

Haupt, K., and Mosbach, K. (1998). Plastic antibodies: developments and applications. *Trends Biotechnol.*, **16**, pp. 468–475.

Hilt, J., and Byrne, M. (2004). Configurational biomimesis in drug delivery molecular imprinting of biologically significant molecules, *Adv. Drug Deliv. Rev.*, **56**, pp. 1599–1620.

Karlsson, J. G., Andersson, L. I., and Nicholls, I. A. (2001). Probing the molecular basis for ligand-selective recognition in molecularly imprinted polymers selective for the local anaesthetic bupivacaine, *Anal. Chim. Acta*, **435**, pp. 57–64.

Kuhn, J., Aylaz, G., Sari, E., Marco, M., Yiu, H. H. P., and Duman, M. (2020). Selective binding of antibiotics using magnetic molecular imprint polymer (MMIP) networks prepared from vinyl-functionalized magnetic nanoparticles, *J. Hazardous Mater.*, **387**, p. 121709.

Liao, Y., Wang, W., and Wang, B. H. (1999). Building fluorescent sensors by template polymerization: the preparation of a fluorescent sensor for l-tryptophan, *Bioorg. Chem.*, **27**, pp. 463–476.

Lee, J., Bernard, S., and Liu, X. C. (2009). Nanostructured biomimetic catalysts for asymmetric hydrogenation of enamides using molecular imprinting technology, *React. Funct. Polym.*, **69**, pp. 650–654.

Lin, H. Y., Hsu, C. Y., Thomas, J. L., Wang, S. E., Chen, H. C., and Chou, T. C. (2006). The microcontact imprinting of proteins: The effect of cross-linking monomers for lysozyme, ribonuclease A and myoglobin. *Biosens. Bioelectron.*, **22**, pp. 534–543.

Liu, X., Wei, Z. H., Huang, Y. P., Yang, J. R., and Liu, Z. S. (2012). Molecularly imprinted nanoparticles with nontailing peaks in capillary electrochromatography, *J. Chromatogr. A*, **1264**, pp. 137–142.

Matsui, J., Kubo, H., and Takeuchi, T. (2000). Molecularly imprinted fluorescent-shift receptors prepared with 2-(trifluoromethyl)acrylic acid, *Anal Chem.*, **72**, pp. 3286–3290.

Mazzotta, E. Malitesta, C. Diaz-Alvarez, M., and Martin-Esteban, A. (2012). Electrosynthesis of molecularly imprinted polypyrrole for the antibiotic levofloxacin, *Thin Solid Films*, **520**, pp. 1938–1943.

Nilsson, K. G. I., Sakaguchi, K., Gemeiner, P., and Mosbach, K. (1995). Molecular imprinting of acetylated carbohydrate derivatives into methacrylic polymers, *J. Chromatogr. A*, **707**, pp. 199–203.

Nilsson, S., Schweitz, L., and Petersson, M. (1997). Three approaches to enantiomer separation of β-adrenergic antagonists by capillary electrochromatography, *Electrophoresis*, **18**, pp. 884–890.

Nostrum, C. F. (2005). Molecular imprinting: a new tool for drug innovation, *Drug Discov. Today Technol.* **2**, pp. 119–124.

Pernites, R., Ponnapati, R. R., and Advincula, R. (2010). Surface plasmon resonance (SPR) detection of theophylline via electro-polymerized molecularly imprinted polythiophenes, *Macromolecules*, **43**, pp. 9724–9735.

Piletsky, S. A., Piletskaya, E. V., Yano, K., Kugimiya, A., Elgersma, A. V., Levi, R., Kahlow, U., Takeuchi, T., Karube, I., Panasyuk T. I., and El'skaya A. V. (1996). A biomimetic receptor system for sialic acid based on molecular imprinting, *Anal. Lett.*, **29**, pp. 157–170.

Piletsky, S. A., Piletska, E. V., Karim, K., Freebairn, K. W., Legge, C. H., and Turner, A. P. F. (2002). polymer cookery: influence of polymerisation conditions on the performance of molecularly imprinted polymers, *Macromolecules*, **35**, pp. 7499–7504.

Polyakov, M. V. (1931). Adsorption properties and structure of silica gel, *Zhur. Fiz. Khim.*, **2**, pp. 799–805.

Puoci, F., Lemma, F., Muzzalupo, R., Spizzirri, U. G., Trombino, S., Cassano, R., and Picci, N. (2004). Spherical molecularly imprinted polymers (SMiPS) via a novel precipitation polymerisation in the controlled delivery of sulfasalazine, *Macromol. Biosci.*, **4**, pp. 22–26.

Puoci, F., Lemma, F., Cirillo, G., Picci, N., Matricardi, P., and Alhaique, F. (2007). Molecularly imprinted polymers for 5-fluorouracil release in biological fluids. *Molecules*, **12**, pp. 805–814.

Puoci, F., Cirillo, G., Curcio, M., Lemma, F., Parisi, O., Spizzirri, U., and Picci. N. (2010). Molecularly imprinted polymers in biomedical applications, in *Biopolymers*, ed., Elnashar, M., Chapter 28 (IntechOpen, London), pp. 547–574.

Sellergren, B., and Allender, C. (2005). Molecularly imprinted polymers: a bridge to advanced drug delivery, *Adv. Drug Deliv. Rev.*, **57**, pp. 1733–1741.

Southard, G. E., Van Houten, K. A., Ott, E. W., and Murray, G. M. (2007). Luminescent sensing of organophosphates using europium(III) containing imprinted polymers prepared by RAFT polymerization, *Anal. Chim. Acta*, **581**, pp. 202–207.

Suda, N., Sunayama, H., Kitayama, Y., Kamon, Y., and Takeuchi T. (2017). Oriented, molecularly imprinted cavities with dual binding sites for highly sensitive and selective recognition of cortisol, *R. Soc. Open Sci.*, **4**, p. 170300.

Tian, S., Guo, Z. F., Zhang, X. X., and Wu, X. Y. (2013). Synthesis of molecularly imprinted co-polymers for recognition of ephedrine, *Anal. Methods*, **5**, pp. 5179–5187.

Venkatesh, S., Sizemore, S. P., and Byrne, M. (2008). Biomimetic hydrogels for enhanced loading and extended release of ocular therapeutics, *Biomaterials*, **28**, pp. 717–724.

Wan, W., Wagner, S., and Rurack, K. (2016). Fluorescent monomers: "Bricks" that make a molecularly imprinted polymer "bright", *Anal. Bioanal. Chem.,* **408**, pp. 1753–1771.

Wulff, G., and Sarhan, A. (1972). Use of polymers with enzymes-analogous structures for the resolution of racemates. *Angew. Chem. Int. Ed.*, **11**, pp. 341–344.

Wulff, G., and Vesper, W. (1978). Preparation of chromatographic sorbents with chiral cavities for racemic resolution, *J. Chromatogr. A*, **167**, pp. 171–186.

Wulff, G. (2002). Enzyme-like catalysis by molecularly imprinted polymers, *Chem Rev.,* **102**, pp. 1–27.

Chapter 6

Bio-Templating: Using Templates from Nature to Form Nanomaterials

Humphrey H. P. Yiu

School of Engineering and Physical Sciences, Heriot-Watt University, Edinburgh, EH14 4AS, UK

h.h.yiu@hw.ac.uk

6.1 Introduction

So far in this book, we have introduced various templating methods for producing a wide range of nanostructured materials, from nanoparticles to polymers with specific cavities at the nanoscale. Their wide range of properties and high chemical/thermal stability allow these nanostructured materials to have applications in a broad spectrum of scientific areas, from biomedical uses to chemical synthesis. Some of these materials have already been made available commercially. However, nature has already been making materials with delicate structures using templating for billions of years. This can be exemplified by commonly found items such as shells, teeth and bones. Indeed, nature provides us with plenty of choices of naturally formed templates for making nanostructured materials. Figure 6.1 shows the delicate structures of marine micro-organisms that could be used for templating

Nanotemplating: Engineering Nanocomposite Materials for the Future
Humphrey H. P. Yiu

ISBN 978-981-4316-92-7 (Hardcover), 978-1-003-50999-8 (eBook)
www.jennystanford.com

purpose. Scientists have already exploited this opportunity to make nanostructured materials with these natural templates.

The idea of "biomimicry" architecture has become popular in recent years in construction. All over the world, you can find many recent construction projects mimicking structures from nature. The idea was that learning from nature may allow a construction with a higher energy- and material-efficiency. It was also believed that such structures can blend into nature more easily, making human life more comfortable (see Figure 6.2). Scientists are also exploring if energy- and material-efficiency can be achieved at a nanoscale using biomimicry.

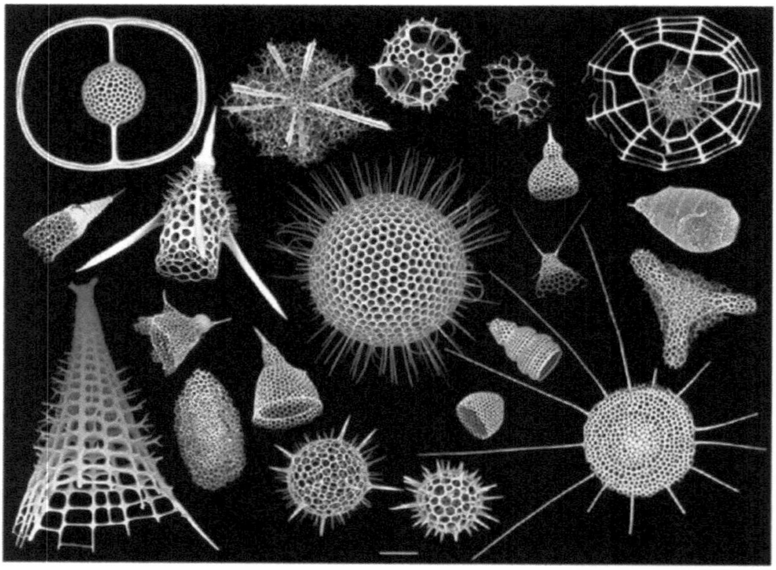

Figure 6.1 Images of silica skeleton of selected marine micro-organisms. *Sources*: Miguel Mendez Sandin, 2018, protocols.io, https://dx.doi.org/10.17504/protocols.io.ug9etz6.

Indeed, in nature, the templating mechanism has been widely adapted by organisms to form required hard structures for various purposes. Organisms tend to utilize precursors that are available naturally to them. For example, calcium ions in seawater are used for constructing the shell (calcium carbonate) for many marine organisms such as mollusks. Some of these hard, inorganic structures are particularly complex in structure, with some in a nanoscale. They can be used as hard bio-templates for preparing nanostructured materials. Also understanding their formation mechanism will help to develop reliable biomineralization routes for synthetic nanomaterials.

Figure 6.2 The idea of biomimicry architecture in modern cities. For example, the (b) Vessel, Hudson Yards, New Your City, resembles the (a) honeycomb structure while (d) Canopy Park, Singapore, looks like (c) the rainforest canopy. *Sources*: (a) Merdal at Turkish Wikipedia, CC BY-SA 3.0, https://commons.wikimedia.org/w/index.php?curid=121469; (b) Stefan Kemmerling, - Own work, CC BY-SA 4.0, https://commons.wikimedia.org/w/index.php?curid=79878351; (c) Rainforest-alliance.org, https://www.rainforest-alliance.org/article/natural-climate-solutions; (d) Sweet life in Singapore, https://sweetlifeinsingapore.wordpress.com/2013/06/.

6.2 Templating Using Micro-Organisms

6.2.1 Crucial Steps in Bio-Templating

The principles of using micro-organisms as templates for producing inorganic nanostructures are similar to that of synthesis for mesoporous inorganics using hard templates. However, the major difference is that these micro-organisms have a defined structure that is unlikely to be altered in the laboratory. Moreover, the composition of the template is also fixed. Therefore, this limits the variety of materials that can be made since template removal has to be carefully considered. Generally, several steps need to be considered; (1) template preparation; (2) deposition of precursors; (3) transformation of precursors (redox, solidification, polymerization, etc.), (4) template removal, if the template is hindering the aimed application(s).

The first stage we need to consider is that if the template requires any pre-treatment before the deposition step. Micro-organisms are of very wide diversity and even more diverse in surface chemistry. More crucially, their behavior could change dramatically when the environment changes. For example, the template micro-organism can be sensitive to pH and temperature. A dead organism may shut down its ion uptake system. Also, in some cases, we only want the hard, inorganic skeleton of the micro-organism as the template, while in other cases, a live organism might be needed for producing the nanomaterials.

The precursor deposition stage can be similar to that for synthesis using a hard template. There are several routes, and the two common ones are simple adsorption on the surface and immersion into a super-saturated solution. Once the precursors are deposited onto the template, transformation might be required. For example, if a metallic nanostructured material is to be made, the metal ion precursor is likely required to be reduced using a reducing agent (electron donor), following the equation: $M^{n+} + ne^- \rightarrow M$. However, some of these inorganic species may not be chemically stable. For instance, a freshly reduced metallic iron can undergo oxidation easily in the presence of air while noble metals, e.g., Au and Pd, are relatively stable in ambient conditions. In case of unstable materials, a passivation step might be needed.

The final obstacle is the template removal stage. As we have seen from previous chapters, conditions required for template removal can be harsh in order to achieve complete removal. Another key consideration is the preservation of the formed nanostructure during template removal. This is particularly difficult and needs careful consideration, allowing the template to dissolve or to be digested while retaining the newly formed inorganic structures. Hence, current available methods are rather limited. These include acid digestion for metal oxides and carbonates (e.g., $CaCO_3$), base digestion for silica, and enzymatic digestion for digestible biopolymers. In some cases, if the template is not interfering with the final application, the template can be left within the nanostructure.

6.2.2 Classic Crystallization Mechanism

Among these steps, precursor deposition and crystallization are among the key steps for the success in forming a robust nanostructure. Understanding the classic crystallization mechanism will help to shed some light on the formation of bio-templating nanostructures. In short, crystallization starts with the formation of clusters from ions (or molecules) in a super-saturated solution and leads to nucleation, followed by crystal growth. In the presence of a template, suitable sites on the surface initiate/facilitate the cluster formation, and then growth of nanoparticles (or crystallites). Various parameters of the solution can control the crystal growth, including the concentration of ions, temperature, pH, counterions, and other structural directing reagents (peptides and polymers). Figure 6.3 shows the stages usually required for bio-templating metal oxides, from crystallization to the final nanostructured product.

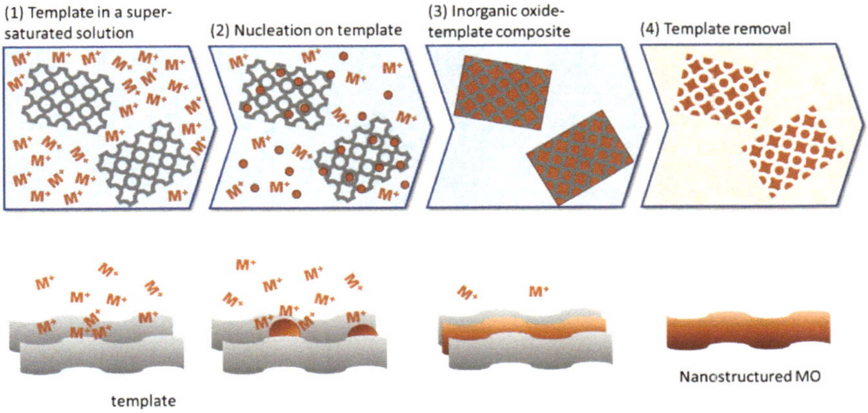

Figure 6.3 Several stages for preparing nanostructured materials using bio-templates.

6.3 Inorganic Nanostructures Templated Using Virus

Viruses are biological species that have a size in the nanoscale, typically in the range of 20–300 nm in size/dimensions. Table 6.1 summarizes the size and shape of selected viruses. Although some viruses cause various deadly diseases to plants, animals and humans, including the recent SARS-CoV-2 virus, many cause no harm to humans due to their narrow "host-range." Therefore, scientists have been exploiting their nanoscale structures to synthesize many inorganic materials with replicated morphologies.

Table 6.1 List of selected viruses with their shapes and sizes

Virus	Shape	Size/diameter (nm)
Ebolavirus	Worm-shape	80 × 970
Tobacco mosaic virus	Tubular	18 × 300
Herpesvirus	Spherical	200
Rabies	"Bullet-shape"	80 × 180
Measles	Spherical	150
HIV-1	Spherical	120
SARS	Spherical	120
Influenza virus	Spherical	100
Dengue virus, Zika virus	Spherical	50
Hepatitis C virus	Spherical	50
Hepatitis B virus	Spherical	42
Hepatitis A virus, poliovirus	Spherical	30

Using viruses as a template for nanostructured materials has been demonstrated in the early 2000s. For example, Dujardin et al. deposited Pt, Au and Ag nanoparticles onto tobacco mosaic viruses (TMV) (Dujardin et al., 2003). It was found that the glutamic and aspartate acid groups on the virus surface provided sites for deposition. This was further supported by Knez et al. (Knez et al., 2004). In the same year, cobalt and nickel nanowires templated using TMV were also reported (Knez et al., 2003).

These nanowires were prepared with TMV activated by Pd^{2+} ions. Interestingly, followed by the deposition of Co or Ni, the nanowire was formed inside the TMV structure. A few years later, a gold (Au) nanowire structure templated from TMV was reported using multiple deposition (addition-reduction cycles) (Bromley et al., 2008). The homogeneity of the deposition was aided by the addition of poly-L-lysine.

Zhou et al. demonstrated the use of TMV as a template for preparing Cu nanorod structures of around 40 nm in width and 1 μm in length (Zhou et al., 2012). These Cu nanorods were formed via activating the TMV surface with Pd deposition (see Fig. 6.4). Their dispersion was also aided by coating with thiolated polyethylene glycol (PEG). In addition to TMV, a similar strategy was also used to make Cu nanorods using bacteriophages *fd* and M13. This showed that such a strategy was not limited to a narrow range of virus species. This group also used conductive polyaniline to induce self-assembly (polymerization) of TMV and coated Pd and Cu over such polyaniline-TMV structure. The polyaniline component further enhances the conductivity of the Pd or Cu nanostructure. This is aimed to form conductive micro-wire at a narrow width of 1–2 μm.

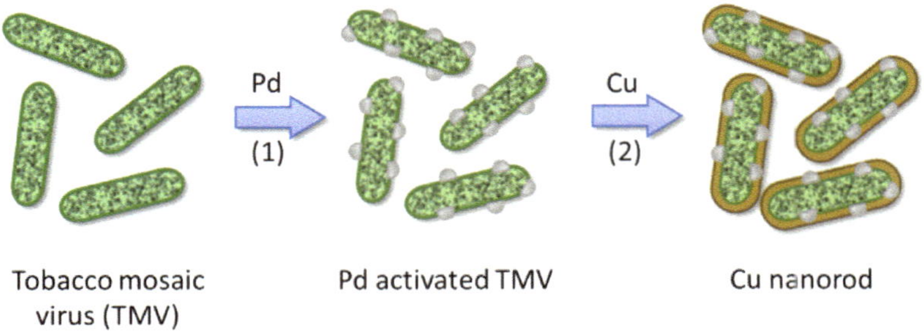

Figure 6.4 Figure illustrating the formation of Cu nanorods from TMV via Pd-activation. Stage (1): deposition of Pd nanoparticles using Na_2PdCl_4, and Stage (2) Cu reductive deposition using $CuSO_4$, dimethylamine borane (reducing agent) and EDTA.

Other materials using TMV as templates include SiO_2 (Royston et al., 2006) and TiO_2 (Rong et al., 2009). More complex TMV-templated structures include Ni-TiO_2 composite nanowires with

Ni as the core coated with TiO_2 (Gerasopoulos et al., 2010). This composite nanowire structure had been tested for use as an anode material, showing high capacity and low average capacity fading in 1000 cycles.

6.4 Inorganic Nanostructures Templated Using Bacteria

In nature, bacteria are thought to be the first form of living organisms on Earth around 4 billion years ago. Most living organisms rely on bacteria for survival since they help to produce many critical chemicals for growth as well as to break down unwanted wastes and also recycle them. Bacteria are also known to be capable of producing nanostructured inorganic particles via biomineralization mechanisms. This could be one major metabolism route but, in some cases, these inorganic materials are essential for the bacterial cells to survive. Therefore, the range of materials that can be produced with this route is limited to those can be found in nature since bacteria are not programmed to make synthetic materials without manipulation from scientists.

6.4.1 Biomineralization of Calcium Carbonate from Bacteria

Calcium carbonate is commonly found in rocks as minerals (calcite and aragonite) and in organisms such as shells and pearls. For bacteria, calcium carbonate is produced from the hydrolysis reaction of urea to form ammonia and carbon dioxide, which is further reacted with calcium ions from the environment to form calcium carbonate polymorphs (see equations below). The enzyme urease (equation 6.1) is required for breaking down urea.

$$CO(NH_2)_2\,(aq) + 2H_2O\,(l) \rightarrow 2NH_4^+\,(aq) + CO_3^{2-}\,(aq) \tag{6.1}$$

$$CO_3^{2-}\,(aq) + Ca^{2+}\,(aq) \rightarrow CaCO_3(s) \tag{6.2}$$

There are many bacteria (e.g., cyanobacteria, *Bacillus*, *Sporosarcina*, *Pseudomonas* and *Azotobacer*) capable in forming calcium carbonate via the biomineralization mechanism (Li et al., 2015). The calcium carbonate formed from this bacteria-induced mechanism tends to have a nano- to sub-micron scaled particle morphology, with spherical or polyhedral crystalline forms. Nanostructured calcium carbonate, in particular nanoporous particles, has been used in drug delivery (e.g., anticancer drug doxorubicin) (Wei et al., 2008). Calcium carbonate-based materials allow drug release when the pH of the environment reduces (to an acidic medium) upon slow dissolution. Since the dissolution products, calcium ions and CO_2, are also not harmful to human bodies, using nanostructured calcium carbonate in biomedicine does seem to be promising. However, research in this area relies on synthetic calcium carbonate nanostructures that possess a highly porous structure for drug storage. The biomineralized calcium carbonate materials from bacterial activity simply may not meet such requirements.

In the literature, biomineralization of calcium carbonate can also be found in "nanobacteria." However, nanobacteria are not considered as living organisms while bacteria are living (Martel et al., 2008). Calcification from nanobacteria was considered to be linked to various serious diseases in humans, from renal and gallbladder stone formations, polycystic kidney diseases and rheumatoid arthritis, to ovarian and nasopharyngeal cancer, as well as Alzheimer's disease. Understanding their mineralization mechanism will help to search for cures for these diseases.

6.4.2 Biomineralization of Calcium Phosphate from Bacteria

Phosphates are widely found in living organisms, e.g., in DNA (nucleicacids and nucleotides), glycerol-phosphate, phosphoproteins and phospholipids. These phosphates are recycled in nature. On the other hand, calcium phosphates, e.g., hydroxyapatite ($Ca_{10}(PO_4)_6(OH)_2$), are strong minerals that can be formed from living organisms as supportive structures. Hence, many structures that require mechanical strength such as skeletons and teeth. Bacteria are also capable of forming calcium phosphate by recycling phosphates from nature. Their formation follows a similar path to calcium carbonate: starting from generating anion (phosphate) from an enzymatic action and then followed by precipitation of an insoluble mineral (calcium phosphate) and mineralization from "seeds" (Follmi, 1996).

One key step in the biomineralization of calcium phosphate is the enzymatic hydrolysis of the phosphoesters from the organism to form the phosphate ions. Enzymes capable of this transformation are phosphatases. Among them, alkaline phosphatase, ALP, is commonly found in bacteria (e.g., *E. coli*) and also in animal liver. Take *E. coli* as an example, *E. coli* (including wild type) can mineralize phosphate via this mechanism. In some cells, internal mineralization (or calcification) also occurs after a week (see Fig. 6.5) (Cosmidis et al., 2015). Although calcium phosphate formed from the bacterial mineralization method is at a nanoscale, the use of such a hard but small material is limited. Again, research in this area tends to be aimed at understanding the formation mechanism which was linked to several diseases in humans. For example, bacteria have been observed on the surface of calcium phosphate kidney stones (Bazin et al., 2012) and calcified cardiac valves (Gilinskaya et al., 2007). Understanding their formation mechanism would help in curing/preventing the disease.

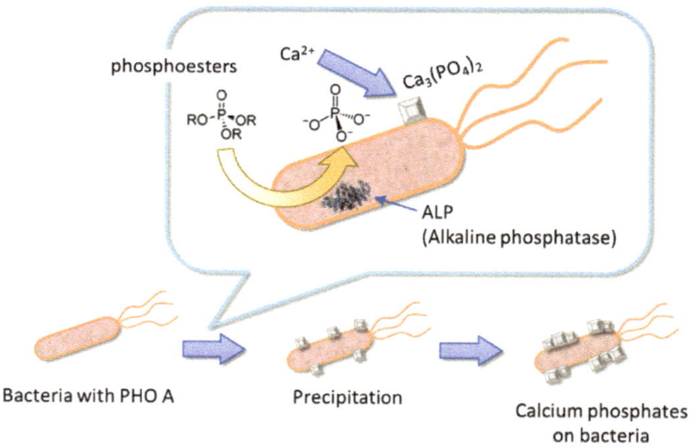

Figure 6.5 Illustration of the biomineralization mechanism of calcium phosphate from bacteria.

6.4.3 Magnetic Nanoparticles from Magnetotactic Bacteria

Among bacteria, there is a group called magnetotactic bacteria, which are capable of producing magnetic nanoparticles (magnetite Fe_3O_4 or greigite Fe_3S_4) inside their cells. These nanoparticles are usually termed magnetosomes, which are coated with a bilayer lipid membrane (similar to a cell structure). In one magnetotactic bacterial cell, 15–20 magnetic nanoparticles of size ranging from 30 to 120 nm, lining up in a chain-like structure. Figure 6.6 shows the structure of these bacteria and their magnetosomes (Amor et al., 2015). Magnetotatic bacteria was first described by Italian scientist, Salvatore Bellini, in 1963 (Bellini, 1963). About 10 years later, Richard Blackmore extensively studied the behavior and characteristics of magnetotactic bacteria (Blakemore, 1975). Many scientists thought this magnetosome helps bacteria to navigate using the Earth's magnetic field, to locate areas with optimal oxygen concentration.

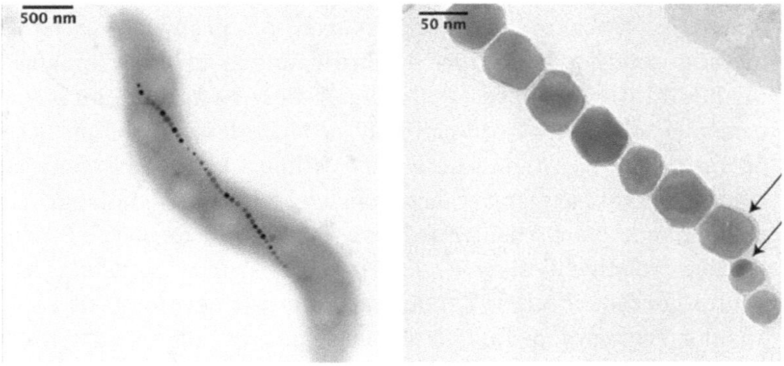

Figure 6.6 (Right) TEM image of a magnetotactic bacteria with a chain of magnetosomes (magnetite nanoparticles) as a line of dark dots. (Left) Extracted magnetosomes with the arrows indicating the membrane coating on the nanoparticles (Amor et al., 2015).

These magnetosome nanoparticles have two main advantages over synthetic magnetite nanoparticles. First is their lipid and peptide coatings, which allow easy functionalization with bio-active molecules such as proteins, enzyme and antibodies. However, many designs of magnetic nanoparticles have already overcome issues with surface chemistry with functionalization that are compatible with biomolecules (see Chapter 1). The second advantage of magnetosomes is their quality in crystallinity. These magnetite particles are usually single domains at a size up to 120 nm, which is rather difficult to prepare with standard precipitation or hydrothermal methods. Single-domain magnetic nanoparticles tend to have a higher magnetic moment, which is usually linked to the domain size. Magnetosome nanoparticles have been used in several biological research areas such as serving as a platform for binding enzymes (Jacob et al., 2020), antibodies (Xu et al., 2019), and genes as a transfection reagent (Dai et al., 2017).

However, there are also two key limitations to using magnetosomes on a large scale. One lies in their formation mechanism. Since the magnetotactic bacteria evolved to generate magnetosomes specifically, it is rather rigid on what materials can be produced. Scientists have tried to substitute iron with other transition metals, including other elements in the same period (Co, Cr, Cu, Ni and Ti) but with little success. Therefore, only two types of nanoparticles can be produced with this method. Another limitation is more common to many biological methods. Ultra large-scale culturing of these bacteria is not always available. Also, every cell can only produce less than 1 picogram of magnetite (Fe_3O_4). Therefore, without large-scale bacterial culture, manufacturing functionalized magnetite nanoparticles using magnetotactic bacteria seems to be impractical. The process is also relatively slower than synthesis from crystallization where good-quality Fe_3O_4 nanoparticles can be produced within an hour. Biological processes can hardly match such speed.

In recent years, scientists have taken this unique behavior of *Magnetospirillum magneticum* one step further. Elfick et al. demonstrated that such behavior in nanoparticle synthesis can be exhibited outside bacteria species (Elfick et al., 2017). This research group transfected human mesenchymal stem cells with the *mms6* gene, which was responsible for the behavior in forming magnetic nanoparticles in magnetotactic bacteria. This could produce magnetic nanoparticles for biomedical uses since these particles would be compatible with human cells. However, there are a few technical issues still to be overcome. First, a rather high iron precursor concentration is required in the cell culture (34 mM of ferric quinate), which could hamper their practicality in large-scale production. Another issue lies on the functionality of the magnetic nanoparticles. Since these particles will be self-generated inside the stem cells, surface functionalization becomes difficult. As we have learnt from Chapter 1, many biomedical applications of magnetic nanoparticles rely heavily on surface functionalities. Nonetheless, this work exhibited the possibility of using biological cells for producing magnetic nanoparticles via biomineralization.

6.5 Inorganic Nanostructures Templated Using Marine Organisms

Biomineralization occurs in many organisms to form various useful structures. The aquatic environment, such as the sea, provides plenty of soluble precursors for the need of bio-mineralization. Therefore, biomineralization of marine organisms has been studied in depth. We can also learn from these processes developing artificial biomineralization for producing useful inorganic nanostructures. Among marine organisms, studies of biomineralization in sponges and diatoms have been widely available in the literature.

6.5.1 Biomineralization from Sponge

Sponges are marine multicellular organisms that are very porous, allowing diffusion of nutrients and circulation of water. For defense purposes, sponges form spicules, which are hard, sharp silica structures. There are many different shapes of spicules, but most of them are well-defined structures with a high degree of symmetry at a microscopic level (see Fig. 6.7). The mechanism for the formation of such delicate inorganic structures is of huge relevance to bio-templating. Along the axial filament of spicules, scientists found that several proteins, silicatein α, β, and γ, of similar structures are responsible for the formation of spicules. Among them, silicatein α belongs to the cysteine protease family. It was found that silicatein α can hydrolyze silicon alkoxides, including tetraethylorthosilicate, to form silica at ambient conditions (Cha et al., 1999). Silicatein α also functions well in a genetically engineered *E. coli* cell (Zhou et al., 1999). However, the exact mechanism of how silicatein α works is still not confirmed. Besides functioning as an enzyme for the hydrolysis of silica sources, silicateins were also found to act as the template for constructing the spicule silica structures. Technically speaking, by immobilizing silicateins on a pre-formed template, microscopic silica structure can be formed (Müller et al., 2009).

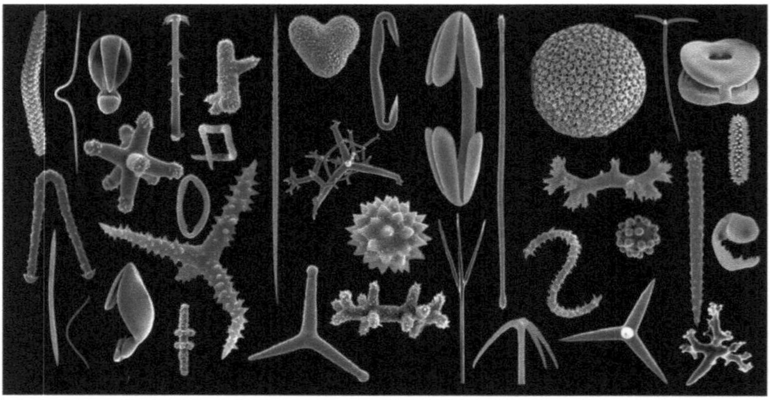

Figure 6.7 SEM images of various spicules found in sponges (not in scale, size varies from 10 μm to 1 mm) (van Soest et al., 2012).

Although nanostructured silicas are of limited use, the formation mechanism of these biosilicas allows the development of other nanostructured oxides which might potentially have more industrial applications. For example, such a mechanism using silicateins has been demonstrated in the formation of nanostructured TiO_2 (Sumerel et al., 2003), Ga_2O_3 (Kisailus et al., 2005), and $BaTiO_3$ (Tao et al., 2010). The last example, nanostructured barium titanate $BaTiO_3$, is of particular interest because barium titanate is considered as a very useful material due to its dielectric, ferroelectric and piezoelectric properties. More interestingly, this mechanism can be used to prepare coated inorganic nanostructured composite. For example, Tahir et al. demonstrated the coating the WS_2 nanotubes with TiO_2 via immobilization of silicatein onto the WS_2 nanotube surface, see

Fig. 6.8 (Tahir et al., 2009). WS_2 nanotube is among the earliest inorganic nanotubes developed after carbon nanotubes, but its inert surface makes it more difficult for functionalization or coating. This opened up a pathway for synthesize a wide range of core-shell materials. However, silicateins are not commercially available and extracting them from sponge is highly laborious and costly. Therefore, commercial production of such material(s) via silicatein biomineralization is unlikely in the near future.

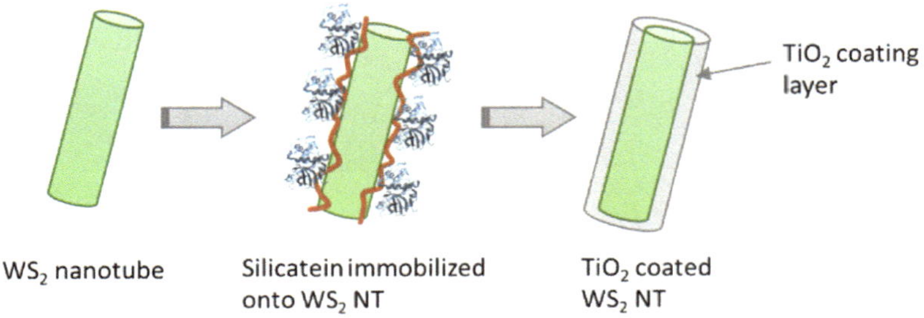

Figure 6.8 Schematic to illustrate how TiO_2 coated WS_2 nanotube can be prepared via silicatein biomineralization.

6.4.2 Biomineralization from Diatoms

Diatoms are one major type of algae found in the aquatic environment globally, including oceans. Algae, including diatoms, are extremely important for our environment and the balance on our planet as they contribute to about 20–50% of the oxygen generation in the world. Diatoms are single-cell organisms with a size in the micrometer range (2–200 μm). Unlike many other micro-organisms, such as bacteria, diatoms have a rigid cell wall called frustule, which is a nanostructured silica material, see Fig. 6.9. Their formation and delicate structures caught scientists' imagination in nano-architecture for useful nanomaterials.

One obvious way to use frustules as bio-templates is to exploit their nanopores for making nanosized materials. This has been demonstrated by Losic et al., who used poly(dimethylsiloxane) (PDMS) to prepare a "stamp" replicate (or negative replicate) from a frustule template *Coscinodiscus sp.* (see Fig. 6.10) (Losic, et al., 2007). With this PDMS template, a positive replicate using a UV curable polymer, NOA 60 (Norland Optical Adhesive 60).

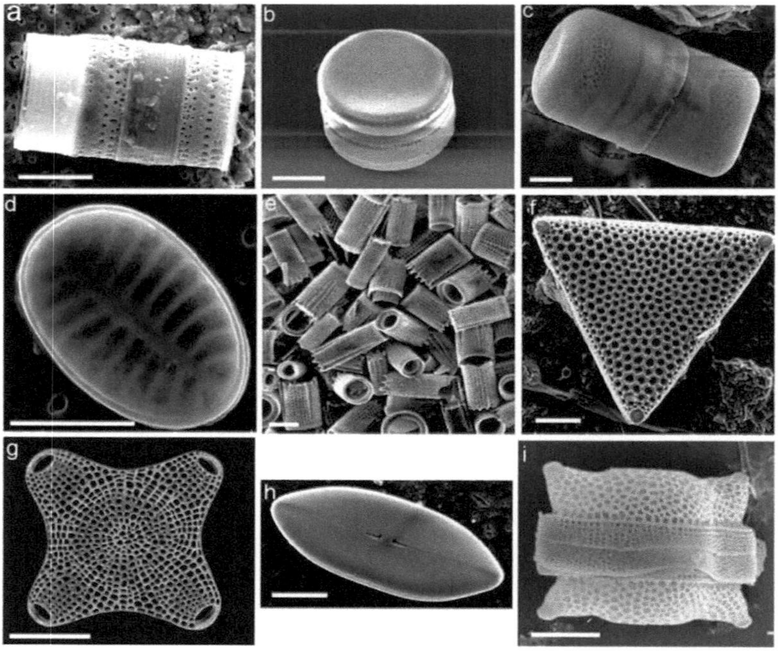

Figure 6.9 SEM images of various frustules from diatoms showing the variety of nano-architecture. Scale bar: 10 μm (De Tammasi et al., 2017).

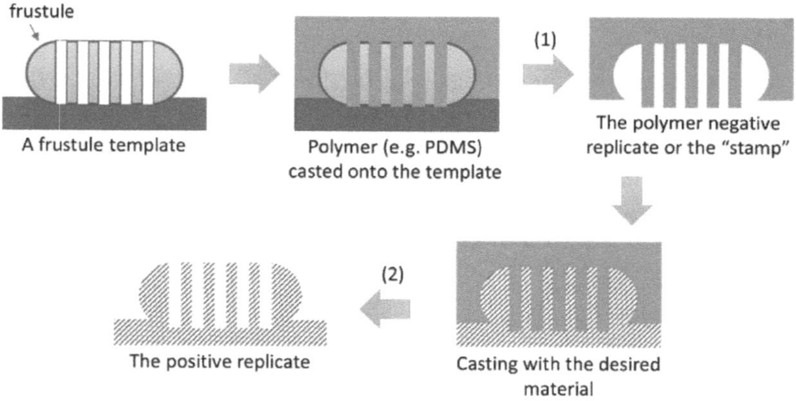

Figure 6.10 Illustration showing the bio-templating route using frustule as a template to form negative and positive replicates. (1) removal of the frustule template using NaOH or HF solvation; (2) removal of polymer negative replicate template using solvent solvation.

This route was similar to that introduced in Chapter 3 for synthesizing the third generation of mesoporous materials, as illustrated in Fig. 6.10. Besides using NOA 60 polymer, the same research group also made a positive replicate of frustule using gold with the same method (deposition followed by evaporation) (Losic et al., 2006). However, the amount of gold precursor used was rather high, making the less likely to be used in industries. Other methods for the deposition of precursors include sputter coating deposition, electrochemical deposition and vapor deposition. It was found that these nanostructured metallic materials enhance fluorescent signals, which finds many applications in biotechnology including biomolecular sensing (Aslan et al., 2005).

Other examples of using frustules as bio-templates include the formation of porous silicon from the silica template (Bao et al., 2007). However, reducing from SiO_2 to elemental Si is not easy and gaseous magnesium Mg is needed (see equation 6.3) at 650°C. Nanostructure silicon materials will be very useful due to their electronic/semiconducting properties. However, this method may not be practical enough for scaling up the production in industry.

$$SiO_2(s) + 2Mg(g) \rightarrow Si(s) + 2MgO(s) \tag{6.3}$$

The same research group also found that by changing the reaction conditions including increasing the temperature to 900°C, a MgO frustule structure was formed while the Si was removed as a Mg-Si liquid. (Sandhage et al., 2002). The same principle, termed BaSIC (Bioclastic and Shapepreserving Inorganic Conversion), has been applied to form other useful materials, including TiO_2, ZrO_2 and $BaTiO_3$ (Nassif et al., 2011).

One advantage of using diatoms for use as bio-templates for nanostructured materials is that algae can be grown commercially on a large scale. This makes scaling up of template production more likely and the cost of templates is not such an obstacle. However, manufacturing using diatoms for nanomaterials still not operated in industries due to other technical problems. For example, the use of NaOH or HF for the removal of the silica frustule templates is still difficult to scale up and possesses a high level of safety concern.

6.6 Applications and Future Outlook

As we can see in this chapter, nature has created many fascinating micro- and nanostructures. We can exploit these structures to produce their replicates using (bio)mineralization methods. Due to the complicated structure on such as small scale, they are difficult to produce by a totally synthetic route. However, also because of this, applications of these bio-templated materials are rather limited. Considering the complicated and laborious procedures just to purify the templates, scaling up the production of some of these templates is virtually impossible. As we highlighted in the example of using magnetotactic bacteria for the production of magnetic nanoparticles (magnetosomes), around 0.1 pg of nanoparticles can be produced from one cell. If we estimated from a standard bacterial culture at 5 m cell/mL, it would take 2000 L to produce only 1 g of nanoparticle, compared to around 200 mL for the co-precipitation method. The amount of culture media and water required to achieve this is simply difficult to justify in an industrial establishment. Therefore, many applications of bio-templating methods are likely to be within academic research, in particular for understanding the biomineralization mechanisms and parameters in nature. To medical research, such understanding is very important for understanding mineralization-related diseases including bone diseases (osteoporosis, osteomalacia, and hypophosphatasia) and other diseases (kidney stones and atherosclerosis). Research in this area can help to search for cures and preventions for these diseases.

Since 1990s, 3D printing has become available, and complicated templates that are considered difficult to make can be printed using a 3D printer. These polymer-based 3D structures can be used for making inorganic replicates. In the late 2000s, 3D-printed structures have also been used in biomedical research and, notably, tissue engineering (Wang et al., 2020). This was made possible with 3D printing using biodegradable polymers. Can this technique be used for producing inorganic nanostructures?

With the current technologies, using 3D printing to mass-produce nanostructured templates is still far from achievable. First, the highest resolution for 3D printing is still only 25 μm (Darkes-Burkey et al., 2020), which is about 100- to 1000-fold improvement still to be required for nanostructures. Another issue is that the polymers used have a certain molecular size, which may not be improved much further for constructing nanostructured templates. However, there can be new technologies that can produce nanostructured templates with controlled patterns for use in nano-templating. Only time can tell.

References

Amor, M., Busigny, V., Durand-Dubief, M., Tharaud, M., Ona-Nguema, G., Gélabert, A., Alphandéry, E., Menguy, N., Benedetti, M. F., Chebbi, I., and Guyot, F. (2015). Chemical signature of magnetotactic bacteria, *Proc. Natl. Acad. Sci. U. S. A.*, **112**, pp. 1699–1703.

Aslan, K., Gryczynski, I., Malicka, J., Matveeva, E., Lakowicz, J. R., and Geddes, C. D. (2005). Metal-enhanced fluorescence: an emerging tool in biotechnology, *Curr. Opin. Biotechnol.*, **16**, pp. 55–62.

Bao, Z., Weatherspoon, M., Shian, S., Cai, Y., Graham, P. D., Allan, S. M., Ahmad, G., Dickerson, M. B., Church, B. C., Kang, Z., Abernathy, H. W. 3rd, Summers, C. J., Liu, M., and Sandhage, K. H. (2007). Chemical reduction of three-dimensional silica micro-assemblies into microporous silicon replicas, *Nature*, **446**, pp. 172–175.

Bazin, D., Daudon, M., Combes, C., and Rey, C. (2012). Characterization and some physicochemical aspects of pathological micro-calcifications, *Chem. Rev.*, **112**, pp. 5092–5120.

Bellini, S. (1963). Su di un particolare comportamento di batteri d'acqua dolce (On a unique behavior of freshwater bacteria), Institute of Microbiology, University of Pavia, Italy.

Blakemore, R. (1975). Magnetotactic Bacteria, *Science*, **190**, pp. 377–379.

Bromley, K. M., Patil, A. J., Perriman, A. W., Stubbs, G., and Mann, S. (2008). Preparation of high quality nanowires by tobacco mosaic virus templating of gold nanoparticles, *J. Mater. Chem.*, **18**, pp. 4796–4801.

Cha, J. N., Shimizu, K., Zhou, Y., Christiansen, S. C., Chmelka, B. F., Stucky, G. D., and Morse, D. E. (1999). Silicatein filaments and subunits from a marine sponge direct the polymerization of silica and silicones *in vitro*, *Proc. Natl. Acad. Sci. U. S. A.*, **96**, 361–365.

Cosmidis, J., Benzerara, K., Guyot, F., Skouri-Panet, F., Duprat, E., Férard, C., Guigner, J.-M., Babonneau, F., and Coelho, C. (2015). Calcium-phosphate biomineralization induced by alkaline phosphatase activity in *Escherichia coli*: localization, kinetics, and potential signatures in the fossil record, *Front. Earth Sci.* **3**, p. 84.

Dai, Q., Long, R., Wang, S., Kankala, R. K., Wang, J., Jiang, W., and Liu, Y. (2017). Bacterial magnetosomes as an efficient gene delivery platform for cancer theranostics. *Microb. Cell Fact.*, **16**, p. 216.

Darkes-Burkey, C., and Shepherd, R. F. (2020). A general fruit acid chelation route for eco-friendly and ambient 3D printing of metals, *Nature*, **588**, pp. 594–595.

De Tommasi, E., Gielis, J., and Rogato, A. (2017). Diatom frustule morphogenesis and function: a multidisciplinary survey, *Mar. Genomics*, **35**, pp. 1–18.

Dujardin, E., Peet, C., Stubbs, G., Culver, J. N., and Mann, S. (2003). Organization of metallic nanoparticles using tobacco mosaic virus templates, *Nano Lett.*, **3**, pp. 413–417.

Elfick, A., Rischitor, G., Mouras, R., Azfer, A., Lungaro, L., Uhlarz, M., Herrmannsdörfer, T., Lucocq, J., Gamal, W., Bagnaninchi, P., Semple, S., and Salter, D. M. (2017). Biosynthesis of magnetic nanoparticles by human mesenchymal stem cells following transfection with the magnetotactic bacterial gene mms6, *Sci. Rep.*, **7**, p. 39755.

Follmi, K. (1996). The phosphorus cycle, phosphogenesis and marine phosphate-rich deposits, *Earth Sci. Rev.*, **40**, pp. 55–124.

Gerasopoulos, K., Chen, X. L., Culver, J., Wang C. S., and Ghodssi, R. (2010). Self-assembled Ni/TiO_2 nanocomposite anodes synthesized via electroless plating and atomic layer deposition on biological scaffolds, *Chem. Commun.*, **46**, pp. 7349–7351.

Gilinskaya, L. G., Zanin, Y. N., and Rudina, N. A. (2007). Bacterial genesis of calcium phosphates in human organism and nature. *Lithol. Miner. Resour.*, **42**, pp. 56–67.

Jacob, J. J., and Suthindhiran, K. (2020). Immobilisation of lipase enzyme onto bacterial magnetosomes for stain removal. *Biotechnol. Rep. (Amst.)*, **17**, p. e00422.

Kisailus, D., Choi, J. H., Weaver, J. C., Yang W. J., and Morse, D. E. (2005). Enzymatic synthesis and nanostructural control of gallium oxide at low temperature, *Adv. Mater.*, **17**, pp. 314–318.

Knez, M., Bittner, A. M., Boes, F., Wege, C., Jeske, H., Maiβ, E., and Kern, K. (2003). Biotemplate synthesis of 3 nm nickel and cobalt nanowires, *Nano Lett.,* **3**, pp. 1079–1082.

Knez, M., Sumser, M., Bittner, A. M., Wege, C., Jeske, H., Martin, T. P., and Kern, K. (2004). Spatially selective nucleation of metal clusters on the tobacco mosaic virus, *Adv. Funct. Mater.,* **14**, pp. 116–124.

Li, X., Chopp, D. L., Russin, W. A., Brannon, P. T., Parsek M. R., and Packman A. I. (2015). Spatial patterns of carbonate biomineralization in biofilms, *Appl. Environ. Microbiol.,* **81**, pp. 7403–7410.

Losic, D., Mitchell, J. G., and Voelcker, N. H. (2006). Fabrication of gold nanostructures by templating from porous diatom frustules, *New J. Chem.,* **30**, pp. 908–914.

Losic, D., Mitchell, J. G., Lal, R., and Voelcker, Dusan, N. H. (2007). Rapid fabrication of micro- and nanoscale patterns by replica molding from diatom biosilica, *Adv. Funct. Mater.,* **17**, pp. 2439–2446.

Martel, J., and Young, J. D. (2008). Purported nanobacteria in human blood as calcium carbonate nanoparticles, *Proc. Natl. Acad. Sci. U. S. A.,* **105**, pp. 5549–5554.

Müller, W. E., Wang, X., Cui, F. Z., Jochum, K. P., Tremel, W., Bill, J., Schröder, H. C., Natalio, F., Schlossmacher, U., and Wiens, M. (2009). Sponge spicules as blueprints for the biofabrication of inorganic–organic composites and biomaterials. *Appl. Microbiol. Biotechnol.,* **83**, pp. 397–413.

Nassif, N., and Livage, J. (2011). From diatoms to silica-based biohybrids. *Chem. Soc. Rev.,* **40**, pp. 849–859.

Rong, J. H., Oberbeck, F., Wang, X. N., Li, X. D., Oxsher, J., Niu, Z. W., and Wang, Q. (2009). Tobacco mosaic virus templated synthesis of one dimensional inorganic–polymer hybrid fibres, *J. Mater. Chem.,* **19**, pp. 2841–2845.

Royston, E., Lee, S. Y., Culver, J. N., and Harris, M. T. (2006). Characterization of silica-coated tobacco mosaic virus, *J. Colloid Interface Sci.,* **298**, pp. 706–712.

Sandhage, K. H., Dickerson, M. B., Huseman, P. M., Caranna, M. A., Clifton, J. D., Bull, T. A., Heibel, T. J., Overton, W. R., and Schoenwaelder, M. E. A. (2002). Novel, bioclastic route to self-assembled, 3D, chemically tailored meso/nanostructures: shape-preserving reactive conversion of biosilica (diatom) microshells, *Adv. Mater.,* **14**, pp. 429–433.

Sumerel, J. L., Yang, W. J., Kisailus, D., Weaver, J. C., Choi J. H., and Morse, D. E. (2003). Biocatalytically templated synthesis of titanium dioxide, *Chem. Mater.,* **15**, pp. 4804–4809.

Tahir, M. N., Natalio, F., Therese, H. A., Yella, A., Metz, N., Shah, M. R., Mugnaioli, E., Berger, R., Theato, P., Schroeder, H. C., Mueller W. E. G., and Tremel, W. (2009). Enzyme-mediated deposition of a TiO_2 coating onto biofunctionalized WS_2 chalcogenide nanotubes, *Adv. Funct. Mater.*, **19**, pp. 285–291.

Tao, A. R., Niesz, K., and Morse, D. E. (2010). Bio-inspired nanofabrication of barium titanate, *J. Mater. Chem.,* **20**, pp. 7916–7923.

Van Soest, R. W., Boury-Esnault, N., Vacelet, J., Dohrmann, M., Erpenbeck, D., De Voogd, N. J., Santodomingo, N., Vanhoorne, B., Kelly, M., and Hooper, J. N. (2012). Global diversity of sponges (Porifera), *PLoS ONE,* **7**, p. e35105.

Wang, C., Huang, W., Zhou, Y., He, L. B., He, Z., Chen, Z. L., He, X., Tian, S., Liao, J. M., Lu, B. H., Wei, Y., and Wang, M. (2020). 3D printing of bone tissue engineering scaffolds, *Bioact. Mater.*, **5**, pp. 82–91.

Wei, W., Ma, G.-H., Hu, G., Yu, D., Mcleish, T., Su Z.-G., and Shen, Z.-Y. (2008). Preparation of hierarchical hollow $CaCO_3$ particles and the application as anticancer drug carrier, *J. Am. Chem. Soc.,* **130**, pp. 15808–15810.

Xu, J., Liu, L., He, J., Ma, S., Li, S., Wang, Z., Xu, T., Jiang, W., Wen, Y., Li, Y., Tian, J., and Li, F. (2019). Engineered magnetosomes fused to functional molecule (protein A) provide a highly effective alternative to commercial immunomagnetic beads, *J. Nanobiotechnol.,* **17**, p. 37.

Zhou, Y., Shimizu, K., Cha, J. N., Stucky, G. D., and Morse, D. E. (1999). Efficient catalysis of polysiloxane synthesis by silicatein α requires specific hydroxy and imidazole functionalities, *Angew. Chem. Int. Ed.,* **38**, pp. 779–782.

Zhou, J. C., Soto, C. M., Chen, M. S., Bruckman, M. A., Moore, M. H., Barry, E., Ratna, B. R., Pehrsson, P. E., Spies, B. R., and Confer, T. S. (2012). Biotemplating rod-like viruses for the synthesis of copper nanorods and nanowires, *J. Nanobiotechnol.,* **10**, p. 18.

Index